OECD Green Growth Studies

Green Growth in Fisheries and Aquaculture

OECD
BETTER POLICIES FOR BETTER LIVES

Please cite this publication as:
OECD (2015), *Green Growth in Fisheries and Aquaculture*, OECD Green Growth Studies, OECD Publishing, Paris.
http://dx.doi.org/10.1787/9789264232143-en

ISBN 978-92-64-23213-6 (print)
ISBN 978-92-64-23214-3 (PDF)

Series: OECD Green Growth Studies
ISSN 2222-9515 (print)
ISSN 2222-9523 (online)

Photo credits: Cover design by Advitam for the OECD.

Corrigenda to OECD publications may be found on line at: *www.oecd.org/about/publishing/corrigenda.htm*.

Foreword

Fisheries and aquaculture depend directly on the natural environment for their productive capacity. The challenges for fisheries and aquaculture policy makers are to deliver on economic and social goals while respecting the natural biological limits of the resource. How to succeed in this has been a central subject of work by the OECD's Committee for Fisheries.

The OECD has been working to help make reforms to boost sustainable growth. At its 50th Anniversary Ministerial Council Meeting in May 2011, the OECD launched a Green Growth Strategy (GGS) to assist policy makers and stakeholders address the major environmental challenges of today's world, while expanding economic opportunities. The OECD GGS provides a set of principles for policy reform and a roadmap to follow that can help to make reform possible, successful and durable. This report combines the sector-specific experience of the COFI with the structure of the OECD GGS in order to provide practical advice to policy makers.

The fisheries and aquaculture sectors face many challenges. Many fisheries are overfished, over-capitalised and suffer from low returns and limited prospects. Aquaculture has grown tremendously, but future growth is threatened by the environmental impacts of production, its dependence on wild fish as feedstock, and competition for space where it operates. To solve these problems, the sectors will have to embrace reform that will put them on the path to greener growth and long-term sustainability.

Fish represents around 20% of the animal protein consumed worldwide and plays an important role in global food security and nutrition. It also disproportionately impacts rural and coastal areas where economic opportunities are limited, especially in less-developed countries. Securing a future with fish and fishing is a necessary part of feeding a growing population and providing inclusive economic opportunities for those who need it most.

The needed changes are possible, and in many places they are happening. The benefits of using the power of markets and trade to put fishing on a sounder economic footing are clear, and the evidence of biological and economic recovery of fish stocks that are managed as part of a sound governance framework is beyond refute. As the number of success stories increases, fisheries around the world are learning from them and adopting new approaches.

The potential of fisheries and aquaculture is impressive. Better management of global fisheries could lead to in excess of USD 50 billion per year in extra profits for fishers, and 13% more fish for consumers. Aquaculture can grow by more than a third over the next ten years if barriers to growth are removed. This report will help to accelerate the process of change by highlighting what works and using the strength of the OECD's evidence-based analysis to produce useful advice.

There is an urgent need to create an enabling policy environment and governance framework for fisheries and aquaculture. But by acting together we can create the momentum for fundamental change.

Acknowledgements

This report has benefitted from the contributions of many authors, as well as the guidance and suggestions of many delegates to the OECD Fisheries Committee. In particular, Kieran Kelleher's work on waste in the fish sector, Claude Menard's work on governance, and the useful insights provided by Josh Abbot are greatly appreciated. Carl-Christian Schmidt was the primary author of Chapter 1, Roger Martini was the author of Chapter 2 and Dong-Sik Woo wrote Chapter 3. Carl-Christian Schmidt supervised the project and provided editorial advice. Roger Martini provided overall editing and was responsible for combining the separate works into a single report. Claire Delpeuch and Lae-Hyung Hong provided comments on earlier versions. Michèle Patterson provided editorial and moral support, and colleagues in the OECD Public Affairs and Communications directorate gave a good deal of advice and assistance in producing the final report. Stefanie Milowski offered invaluable administrative support.

Acronyms

AIMAP	Aquaculture Innovation and Market Access Programme (Canada)
CBFM	Community-Based Fisheries Management
CCTV	Closed-Circuit Television
CFP	Common Fisheries Policy of the European Union
COFI	Committee for Fisheries of the OECD
CPUE	Catch per Unit Effort
CQM	Catch Quota Management
DFO	Fisheries and Oceans Canada (Federal Government)
EEZ	Exclusive Economic Zone (typically seas within 200 nautical miles of coast)
EPA	Environmental Protection Agency (United States)
FAO	Food and Agriculture Organization of the United Nations
FCR	Feed Conversion Ratio
FTC	Fuel Tax Concessions
GFT	Government Financial Transfer (OECD Database)
GGS	OECD Green Growth Strategy
GHG	Greenhouse Gasses
HCR	Harvest Control Rule
ICES	International Council for the Exploration of the Sea
IFFO	International Fishmeal and Fish Oil Organisation
IOM	Integrated Ocean Management
ITQ	Individual Transferrable Quota
IUU	Illegal, Unreported, and Unregulated Fishing
MCM	Meeting of the OECD Council at the Ministerial Level
MCS	Monitoring, Control and Surveillance
MEY	Maximum Economic Yield; the harvest rate that maximises fishing profits over time
MSA	Magnuson-Stevens Act; main federal law concerning fishing in the United States
MSY	Maximum Sustainable Yield; the harvest rate that maximises the quantity of fish over time
NGO	Non-Governmental Organisation
NMFS	National Marine Fisheries Service. An agency of NOAA with responsibility for fisheries (United States)
NOAA	National Oceanographic and Atmospheric Administration (United States)
NRDC	Natural Resources Defence Council. An NGO.
RAC	Regional Advisory Council (European Union)
RAS	Recirculating Aquaculture System
REM	Remote Electronic Monitoring
RFMC	Regional Fisheries Management Council (United States)

RFMO	Regional Fisheries Management Organisation.
TAC	Total Allowable Catch
TURF	Territorial use rights in fisheries
UNCLOS	United Nations Convention on the Law of the Sea
UNEP	United Nations Environmental Programme
UNGA	General Assembly of the United Nations

Table of contents

Acronyms 5

Executive summary 11

Chapter 1. Green growth in the blue economy: Integrating fisheries, aquaculture and the environment 13

Overview of the fisheries and aquaculture sectors 14
The OECD green growth strategy 18
References 34

Chapter 2. Green growth in fisheries 37

Build a green growth model 38
Promote the transition 46
Enable change 61
Measure progress towards objectives 69
Deliver on green growth 70
References 74

Chapter 3. Green growth in aquaculture 79

Aquaculture: A history of rapid growth, with concerns about sustainability 80
Build a green growth model in aquaculture 83
Enable change 100
Deliver on green growth 105
Key findings 108
References 110

Tables

Table 1.1. Green Growth Policy Toolkit: Some possible examples in fisheries 20
Table 1.2. Typology of fishery service delivery and payment models 25
Table 1.3. Some potential green growth indicators in fisheries 31
Table 2.1. Potential gains from reducing waste in Fisheries 44
Table 2.2. Elements of good governance 52
Table 3.1. Major species cultured in OECD countries in 2011 81
Table 3.2. Aquaculture volume and value, 1970-2010 82
Table 3.3. Green Growth challenges and aquaculture 84
Table 3.4. Policy tools to address externalities 87
Table 3.5. Comparison of sustainability indicators for animal protein production system 90
Table 3.6. Calculation of pelagic forage fish equivalent per unit of cultured species groups 97
Table 3.7. Impact of production of from four aquaculture systems 99
Table 3.8. Input use for salmon farming and salmon feed milling in Norway, the United Kingdom, Canada and Chile, 2007 100

Figures

Figure 1.1. Exploitation status of fisheries 14
Figure 1.2. Fish from capture or aquaculture, 1986-2022 15
Figure 1.3. Moving from single species management to a 18
Figure 1.4. The OECD green growth strategy 19
Figure 1.5. A concept of sustainability indicators in Norway 33
Figure 2.1. Risks of business as usual in fisheries 41
Figure 2.2. Historical progression of fisheries management 42
Figure 2.3. Number of fishers and fishing vessels, selected OECD countries 2001-2011 43
Figure 2.4. International Fisheries Governance Structure 45
Figure 2.5. Green growth diagnostic 47
Figure 2.6. An example of a delegated system 50
Figure 2.7. MSY vs. MEY 60
Figure 2.8. Three possible recovery trajectories 65
Figure 3.1. . Aquaculture production by type 82
Figure 3.2. Total aquaculture production: Volume and value 83
Figure 3.3. World fish meal and fish oil production from 1976 to 2006 97
Figure 3.4. Fishmeal use: Past projections and current trends 98
Figure 3.5. Estimated global use of fish meal and oil by the salmon farming industry projected to 2020 98
Figure 3.6. Use of antibiotics in the Norwegian aquaculture industry 103
Figure 3.7. Danish aquaculture production between 1980 and 2010 106
Figure 3.8. Atlantic salmon production among major countries 106

Boxes

Box 1.1. Sustaining growth in the marine and maritime sectors 18
Box 1.2. Green growth and sustainable development 21
Box 1.3. Market-based instruments in Denmark 24
Box 1.4. Cost recovery in fisheries 24
Box 1.5. Recreational vs. commercial fishing in the United States 26
Box 1.6. Innovation for Green Growth 26
Box 1.7. Danish experiments with monitoring and remote sensing 27
Box 1.8. The use of fuel flow meters in the United Kingdom 28
Box 1.9. Canadian Aquaculture Innovation and Market Access Programme (2008-2013) 29
Box 1.10. Headline indicators 30
Box 1.11. Key principles in selecting indicators to monitor progress in green growth 30
Box 2.1. The Aichi Strategic Goal B Target 6 38
Box 2.2. What is IUU fishing? 40
Box 2.3. Addressing overcapacity in the European Union 40
Box 2.4. Discards waste resources and complicate management 41
Box 2.5. A brief history of fisheries management 48
Box 2.6. Community-Based Fisheries Management in Korea 51
Box 2.7. Advisory Councils in the European Union 51
Box 2.8. Rebuilding under the US Magnuson Stevens Act 53
Box 2.9. US regulations set a basis for reduction of food waste and hunger 57
Box 2.10. The equimarginal principle and its shortcomings 59
Box 2.11. Preserving small communities in the United Dingdom 62
Box 2.12. Principles and guidelines for decommissioning schemes 63
Box 2.13. Impact of quota reduction in the US Northeast 66
Box 2.14. Scallops in the Bay of Saint-Brieuc 67

Box 2.15. Initial allocations 68
Box 2.16. Iceland's experience with ITQs 69
Box 2.17. What have we learned from attempts to introduce green-growth policies? 71
Box 3.1. Optimal regulation in aquaculture 86
Box 3.2. Aquaculture of filter-feeding molluscs 90
Box 3.3. Resolving site conflicts: The case of Turkey 91
Box 3.4. The recovery of the Chilean salmon industry 93
Box 3.5. The revival of tiger shrimp industry from white spot disease: The Sri Lanka case 94
Box 3.6. Moving towards a zoning structure in the Norwegian aquaculture 94
Box 3.7. Dealing with externalities from other sectors: the French experience 95
Box 3.8. Seaweed aquaculture for green growth in Korea 101
Box 3.9. Strategic guidelines for the sustainable development of EU aquaculture 101
Box 3.10. Gender inequality issues in aquaculture 102
Box 3.11. Innovations strategies and green growth: The case of Norwegian salmon farming 104
Box 3.12. Strategic guidelines for the sustainable development of EU aquaculture 107

Executive summary

A systematic and inclusive approach to green growth

Green growth means fostering economic growth and development while ensuring that natural assets continue to provide the resources and environmental services on which our well-being relies. The OECD green growth strategy sets out a five-step roadmap to establish sustainable pro-growth reforms. This begins with preparing the way for change, removing barriers to reform, establishing a new policy set and finally by measuring results and taking stock. Following this roadmap means that policies for the fisheries and aquaculture sectors must be more systematic in identifying risk, using market-based instruments and building accountable institutions. It also calls for a more expansive process, involving a broad reach of communities dependent on the fisheries and aquaculture sectors. This will, *inter alia*, require investing in better data and science to support managing a more complex system while building networks to ensure policy coherence and inclusiveness.

Fisheries need reform towards efficient, effective and inclusive management

The practical tasks to enable green growth are to rebuild fisheries to their full potential, ensure that management systems are effective at controlling fishing and to maximise the value of the fishery to fishers and society. Primary attention should be paid to stock management as the cornerstone of sustainable use. In addition, managing the quality of the ecosystem (biodiversity, habitat, and pollution) is essential for ensuring that the resource and its habitat at large are as productive as possible. The increasingly diverse and intensive use of marine space necessitates mitigating spill-over impacts between users and ensuring policy coherence across domains through a coherent and inclusive approach to marine resources management.

There are a number of ways in which fisheries can grow while respecting the natural limits of the ecosystem. It has been estimated that fisheries could deliver 13% more fish in volume and an additional USD 50 billion or more per year in profits if all depleted stocks were recovered and efficiently managed. Moreover, waste in the form of discards at sea and processing losses can make up as much as 40% of the harvest in some fisheries. Eliminating the "race to fish" through better management can significantly improve the energy efficiency of fishing, reducing its carbon footprint and costs for fishers.

Aquaculture needs an enabling environment that addresses externalities

Growth in aquaculture has been nothing short of remarkable, transforming the sector from a niche to the most important source of fish products for consumers today. The challenge for aquaculture is to maintain growth in the face of environmental, spatial or legal limits. This will require innovative policy solutions and a broader stakeholder acceptance of the future role of aquaculture. Future aquaculture will produce more fish in smaller spaces, be less dependent on wild feedstock for feed, emit fewer pollutants, and mix better with other coastal activities. This will come from innovative new technologies, combined with enabling conditions that identify new solutions to satisfy a range of user groups dependent on the marine environment.

In OECD countries in particular, the regulatory landscape is an important factor for future growth. National development plans, institutional innovation, certification, and spatial planning have all been identified as ways to improve the prospects of aquaculture. In particular, market-based approaches can help ensure that the most efficient producers have the opportunity to expand, that marine spaces are allocated to their best possible use, and that aquaculture is attractive for investments in production and innovation.

Future success requires a broader perspective

The common thread in these otherwise very different sectors is the need to plan in a way that encompasses the larger marine economy and environment. This requires investing in both science and the capacity to use it effectively in policy making. Developing good indicators are an important way to convert scientific information into a usable form.

Policy directions

- A whole-of-government approach to broad policy setting should be taken, using clear and measurable benchmarks for success and treating green growth as an opportunity to continually evaluate and improve policies. In particular, policy makers should ensure that there is coherence between fisheries and aquaculture policies.
- Institutions of fisheries management should maintain a disciplined approach that puts science and data first to manage stocks sustainably. As our understanding of the marine environment grows, ecosystem considerations should be incorporated into stock management.
- Improved efficiency is the only way to drive continued growth despite the natural biological limits of the fisheries resource. This demands that all aspects of the sector, both public and private, be innovative and dynamic. Policies to improve productivity and resiliency of fisheries and aquaculture can also help push out those limits over time.
- Governments should embrace their role as facilitators of green growth by removing barriers to change, supporting innovation, and using the power of markets to help maximise profitability.
- Involving all stakeholders can yield more effective and informed policies, improve implementation and create incentives for self- enforcement. Achieving this in practice requires clearly defined roles and processes for appeal of decisions.
- Marine spatial planning approaches can reduce uncertainty, facilitate investment, take into account environmental concerns, and help avoid conflicts among users.
- Reliable statistics are important for good management. Governments should ensure that their data collection is cost-effective and relevant.
- In parallel with improved statistics, effective monitoring and evaluation frameworks need to be developed to ensure that actions are taken and that they lead to concrete benefits. Lessons learned can help other countries still considering green growth strategies (sharing experience) and can help create best practices.
- Sharing best practices in aquaculture rearing processes or policies or institutional arrangements internationally can help both individual farms and countries moving toward green growth. Special assistance might be appropriate in this regard for developing countries.
- More effective and efficient regulations that provide the right incentives can lead to growth. Effectively addressing environmental externalities is key to unlocking future growth potential.

Chapter 1

Green Growth in the blue economy: Integrating fisheries, aquaculture and the environment

This chapter describes the fisheries and aquaculture sectors in broad terms, identifying some of their important interactions and shared challenges for the future. It shows that the prospects for sustained growth are good if reforms along the lines suggested by the OECD Green Growth Strategy are undertaken. Well-managed fisheries can deliver billions more in value and millions of tonnes more fish each year, while aquaculture has the potential for continued strong growth to supply the food requirements of a growing world. An integrated policy view that takes in the whole of the "blue economy" of marine and coastal spaces and which evaluates itself against clear and measureable objectives is required to secure this vision of green growth for fisheries and aquaculture.

Overview of the fisheries and aquaculture sectors

Global demand for fish products has increased over the last decades and this trend is expected to continue due to population growth and increasing wealth, as well as a growing preference for healthy foods (Garcia and Rosenberg, 2010). Given the natural limits to capture fisheries production, it is clear that aquaculture will have to meet most of the future increase in demand for fish (Bostock et al., 2010) (Figure 1.1). In 2021, world fisheries and aquaculture production is projected to be about 172 million tonnes, which is a 15% increase from the average level for 2009-11. The OECD-FAO Agriculture Outlook projects that by 2023 (compared to a baseline of 2013) aquaculture will grow by 38% compared to 2% growth of capture fisheries (OECD/FAO, 2014).

Almost everywhere people are concerned about the sustainability of fishing. In many places, this is for good reason. Current evidence indicates that more than a quarter of fish stocks are overfished, with the rest increasingly being fished at their maximum capacity (Figure 1.1). This indicates that the pressure placed on fish stocks is significant, and sometimes excessive. Not only fish stocks are at risk. Harmful fishing practices can adversely impact aquatic environments and harm non-targeted species. This adverse impact to the overall marine environment is troubling for many reasons, but it also ultimately reduces the capacity of the ocean to produce the fish sought after by fishers.

Figure 1.1. Exploitation status of fisheries

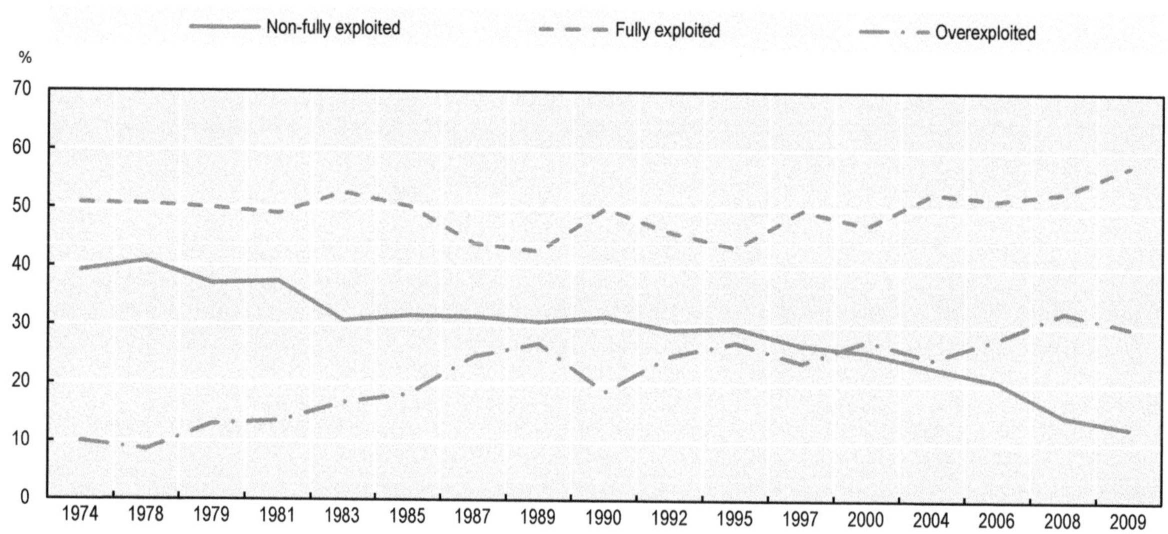

Source: FAO (2012), *State of the world's fisheries*, Food and Agriculture Organization Publications, Rome.

Over the past decades many countries have made great strides in rebuilding their fish stocks and establishing robust management regimes, though much remains to be done. At the international level there has been a range of activities that span from the setting-up of a number of regional fisheries management organisations (RFMOs) to the adoption of new hard and soft law. Examples include the 1995 Fish Stocks Agreements, the FAO Code of Conduct and the WSSD Johannesburg 2002 Plan of Action. In addition a number of private initiatives – including labelling initiatives – have contributed significantly to heighten awareness that action is needed to arrest overfishing and to preserve and share the environmental assets that underlie the fishery.

It has been estimated that if global fisheries were optimally managed, they would generate an additional USD 50 billion in extra income or more (World Bank, 2009; Sumaila et al., 2012) and could produce 13% more fish by 2030 (World Bank, 2013). So, despite the fact that fisheries do have an ultimate productive limit, we are currently at some distance from that point. Significant growth in the value of fisheries is possible without putting additional strain on the resource. All that is needed is more effective management targeting economically optimal stock levels.

As a percentage of GDP, among OECD countries fisheries is a significant industry only in Iceland and New Zealand. But its role in remote and rural areas can be very important in terms of jobs and economic activity such as in Japan, Canada and Norway. Outside of the OECD, it is an important source of nutrition and employment, especially in coastal areas of developing States and territories. But in terms of total food supply, fishing is set to be outstripped by aquaculture (Figure 1.2).

Figure 1.2. Fish from capture or aquaculture, 1986-2022

Thousand tonnes

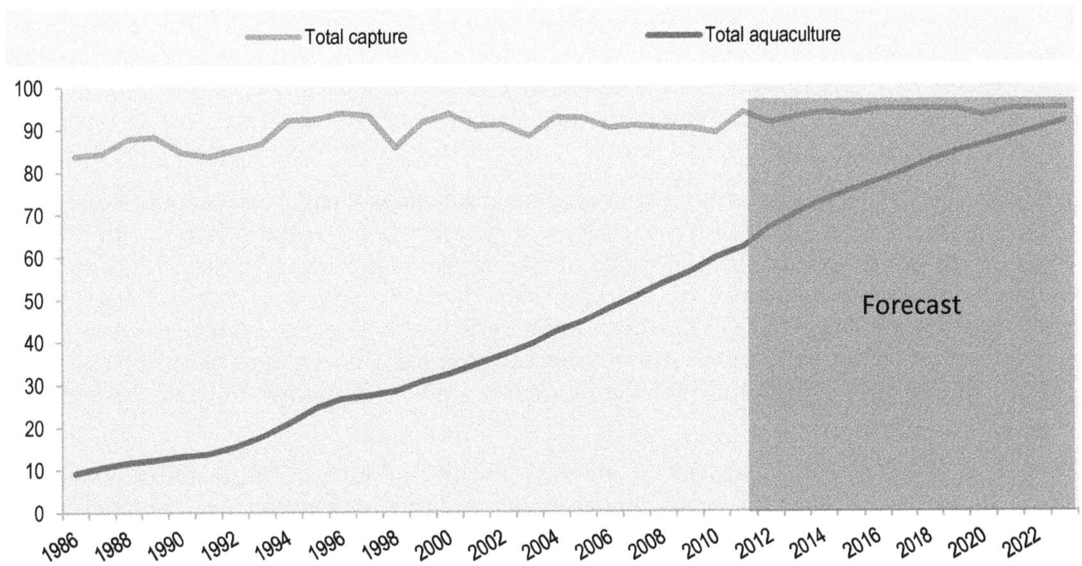

Source: OECD/FAO (2012), *OECD-FAO Agricultural Outlook 2012*, OECD Publishing, Paris. DOI: http://dx.doi.org/10.1787/agr_outlook-2012-en.

Aquaculture grew at an average annual rate of 8.4% between 1970 and 2009. It has been one of the fastest growing food producing sectors in the world, and its potential to contribute to the global food supply is significant. In 2009, aquaculture contributed 38% to the world's fisheries production (excluding aquatic plants) and contributed to about half of all seafood consumed by humans (FAO, 2011).

Aquaculture has potentially a major role in helping to reduce poverty and increase foreign currency earnings, especially for countries in development. Increased production, together with innovation in aquaculture, has lowered production costs significantly and has provided benefits to consumers and producers. For example, shrimp production increased 43 times (from 72 000 tonnes to 3.1 million tonnes) between 1984 and 2007. Concurrently, the price decreased to less than half of what it was originally (from USD 16.40 per kilo to USD 7 per kilo) (Asche, 2008).

Fisheries and aquaculture interactions

The capture fisheries and aquaculture sectors intersect in a number of important ways. Fishmeal and oil from capture fisheries are used in feed compounds; marine space is needed by both activities and their products compete in the marketplace. The line between wild and cultured stocks is sometimes blurred by the use of restocking from hatcheries or the ranching of wild species.

Aquaculture and capture fisheries production share a common market for fish and fish products. While capture fish may carry a premium (rarity, size and quality differences) farmed fish has several other advantages including standardised size, year-round availability and a known production environment.

Policy helps determine the relationship between capture fisheries and aquaculture, and policy-makers face difficult choices. Should the available space be used for capture fishing or aquaculture? Should the fish be used as food fish or in feed compounds for aquaculture? And are capture fish more or less "green" or environmentally responsible than their farmed cousins? Support directed to either sector can impact competitiveness and the balance between the two.

Capture fisheries supply feed for aquaculture

Most carnivorous fish need some amount of fishmeal and fish oil in their diets. The global production of fishmeal and fish oil has stabilised around 6 to 7 million tonnes of fishmeal and 1 million tonnes of fish oil per year. In 2010, 73% of fishmeal and 71% of fish oil was used by aquaculture. The cost and availability of fishmeal and oil is a challenge to future aquaculture growth. According to the International Fishmeal and Fish Oil Organisation replacement of fishmeal and oil is an on-going issue mainly driven by the cost of fishmeal and oil as an ingredient in the feed compounds (IFFO, 2012).

The share of fishmeal in the overall feed mix has declined as alternative sources (mainly soybean meal) are introduced. Fish oil continues to be a key ingredient in aquaculture feed due to its Omega 3 content. Fish oil is also used in the high-value human dietary supplements market. IFFO believes that aquaculture can continue to reduce its use of fish oil but only within certain technical limits. Oil produced from marine algae and genetically-modified grains may eventually come to some markets in some countries, replacing some of the fish oil presently sourced from capture fisheries and trimmings. Fish with high Omega-3 content may ultimately become a premium product to those fed with feed not containing this nutrient.

Most fish used for fish meal and oil are not suitable for human consumption, though this is changing. Increasing demand for aquaculture feeds must not undermine the management of the "industrial" fish used to produce them. This includes ensuring that the ecological role of these species as prey fish is maintained (some fish may be more valuable as prey for wild species than as feed) and that the fishery is soundly managed. To help support this, IFFO has developed a "Global Standard for Responsible Supply" (GSRS) based on the FAO Code of Conduct for Responsible Fisheries.

Governments can act to promote innovations to reduce the use of fish meal and oil in feed compounds for carnivorous farmed species, must ensure, through certification or other means that the fish stocks used for fish meal and oil products are sustainably managed. Promoting non-carnivorous aquaculture species can weaken the link between capture and aquaculture. Species like catfish, carps and *pangasius* require little or no use of fishmeal and oil. Currently, however, in most OECD markets farmed species like salmon, turbot, sea bass and bream have higher consumer appeal.

Fisheries and aquaculture share marine spaces

Competition for marine space increases with the number and extent of economic activities that depends on it. Beyond capture fisheries and aquaculture, tourism, maritime transport, extraction of minerals and hydrocarbons, production of electricity through windmills and tidal wave systems, naval activity, dumping and disposal of waste from production on land all use marine space. In addition to space conflicts these activities will also compete for scarce resources in harbours and other infrastructure.

Ideally, allocation of access or user rights to marine space would be based on the added value to society of the activity. Traditional approaches to regulation can work, but designing and developing institutions in such a way that different users can negotiate and compete directly will also be important. This reduces the burden on the regulator to measure and evaluate the relative benefits of different uses, thereby lowering costs while increasing efficiency in allocation.

Policy approaches such as integrated ocean management have a number of ingredients (notably multi-sectoral management and ecosystem-based management) and component tools (such as marine spatial planning), and should in particular include fisheries management. Boundaries, scope and scale, the range of instruments and institutions for IOM, including ocean zoning, rights-based approaches, and governance institutions and the role of economics in design and implementation of integrated ocean management are all issues that need to be dealt with (Charles, 2011).

Some fish are both wild and farmed

Depleted fish stocks can in some cases be rebuilt by enhancements that can restore fishing possibilities for commercial or recreational fishers. There are three types of enhancement related to fisheries; restocking refers to the release of cultured juvenile fish into the wild in order to restore or re-establish a severely depleted spawning biomass; stock enhancement is the release of cultured juveniles into the wild to augment the natural supply of juveniles to optimise harvests; and sea ranching refers to the release of cultured juveniles into unenclosed marine and estuarine environments for harvest at a larger size in "put, grow, and take" operations (Bell et al., 2008).

While not always effective, sometimes fisheries enhancement is the only means to replenish a water body with fish or other aquatic animals. The root causes of the stock decline must be addressed, whether it is overfishing, dam building, pollution or something else, so that the replenished stock does not meet the same fate. Restocking can lead to weaker genetic material (a problem known in particular for salmon, which have unique genetics for each spawning group.

Catching juveniles for fattening in pens has become an important practice, especially for Bluefin tuna. In essence fish farmers appropriate a wild resource and grow them to marketable size. The practice allows the fish farmer to add value to a wild harvest by feeding lower value fish to larger-bodied high value carnivorous fish. This is a form of aquaculture that relies particularly strongly on wild stocks, and in particular juvenile stocks, such that proper accounting of its impact as part of the management of the wild stock is especially important.

Broadening the scope of policy-making

The activities of the fisheries and aquaculture sectors do not take place in a vacuum. Feed and competition for space are particular important, but also fisheries and aquaculture activities compete with tourism, shipping, exploration for hydrocarbons, agriculture (for land-based aquaculture and for run-off in the coastal zone) electricity generation and other activities.

Fisheries and aquaculture are also more broadly connected with the ecosystem and its different users (Figure 1.3). Going beyond stock management to include broader objectives can bring the coherence between policy domains that is a key element of the GGS. Taking a broader view and operating at a higher level also comes at a higher cost. Additional efforts are required to ensure policy coherence and co-operation across policy domains that affect the ocean activities, including for data and information, surveillance, enforcement and administration.

Management of individual species of fish stocks is usually handled by a single institution, with stakeholder involvement limited to user groups with a direct interest in the fishing sector. When managing at higher levels, i.e. ecosystems or integrated ocean management, interest groups are spread more widely and pursue a variety of economic activities and more ministries and agencies have competence. For example, managing coastal eco-systems might require agriculture, rural development, fisheries, aquaculture, tourist, zoning interests etc. to be taken into account in decision making. This complexity increases costs, adds uncertainty and time to the process, and as well introduces information asymmetries. This needs to be weighed against the benefits of improved management.

Figure 1.3. Moving from single species management to a "blue economy" approach

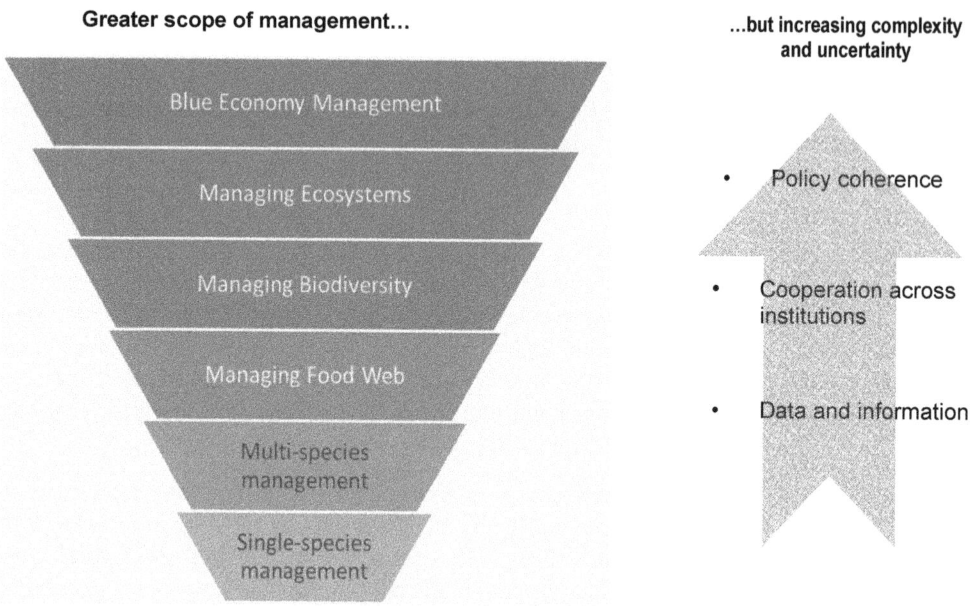

Most ocean policies aspire to take a broader approach and recognise that this is key to maximising benefits (Box 1.1). In practice however most countries find it prudent to stick to a single-species approach to management. The OECD's Green Growth Strategy calls for establishing more effective institutional arrangements that can help countries to eventually take a more holistic approach. However, the complexity of the marine ecosystem tests the limits of science and knowledge. Moving to more complex management models will not be done overnight.

Box 1.1. Sustaining growth in the marine and maritime sectors

The European Union's long-term strategy to support sustainable growth in the marine and maritime sectors as a whole is called Blue Growth. It recognises that seas and oceans are drivers for the European economy with great potential for innovation and growth. An analysis of the job-creation potential, as well as the potential for research and development, has suggested that the following five value chains could deliver sustainable growth and jobs in the blue economy: ocean energy, aquaculture, blue biotechnology, coastal tourism and marine mineral extraction. On-going EU initiatives are already encouraging innovation in sectors such as maritime transport. Other value chains may emerge over time as suitable areas for further policy focus. Blue growth is meant to contribute to sustainable growth and it will need to be coherent with the principles of green growth.

Source: European Commission.

The OECD green growth strategy

The OECD Green Growth Strategy outlines policy principles to help countries ensure that they can obtain the highest possible level of economic growth while conserving the natural resource base upon which that growth depends. The objective of the GGS is hard to disagree with - do the best you can while protecting the environment and social values. The value of the GGS lies not in its objectives, but in the specific approach it proposes and the principles that underpin it.

Policies that support green growth should lead to outcomes where either resource use per unit of value added declines (called relative decoupling) or resource use and environmental impacts are kept stable or declining while the economy is growing overall (absolute decoupling). In fisheries, this means that the primary resource (fish stocks) should be managed at some optimum level while the economic value delivered from that resource is continually increased. That increased value could come from larger harvests supported by a healthier stock and ecosystem, from innovation that adds value to fish products or from productivity improvements that reduce costs, especially from reducing overcapacity.

For fisheries, the main tasks to enable green growth are to rebuild fisheries that are below their potential, ensure that management systems are effective at controlling fishing and to maximise the potential value of the fishery to society. Primary attention should be paid to stock management as the cornerstone of sustainable use. In addition, maintaining the quality of the ecosystem (biodiversity, habitat, and pollution) is essential for ensuring that the resource and its habitat at large are as productive as possible. Marine spaces are increasingly under more diverse and intensive use, such that mitigating spill-over impacts on other users, and vice-versa, is making policy coherence across domains an important consideration.

The aquaculture industry is increasingly seen as an opportunity to improve food security and promote growth. The aquaculture sector will rely on green growth principles to ensure that its rapid growth is sustainable into the future. That is, the aquaculture industry must find a way to provide more seafood to meet increasing demand while mitigating the potentially environmentally negative effects of production.

The OECD GGS is a practical tool that helps policy makers establish the conditions for success in designing their own strategies for growth. The OECD GGS identifies a number of specific steps that must be taken to improve growth and welfare while enhancing sustainability. For each of these steps, specific actions can be envisaged (Figure 1.4).

Figure 1.4. The OECD green growth strategy

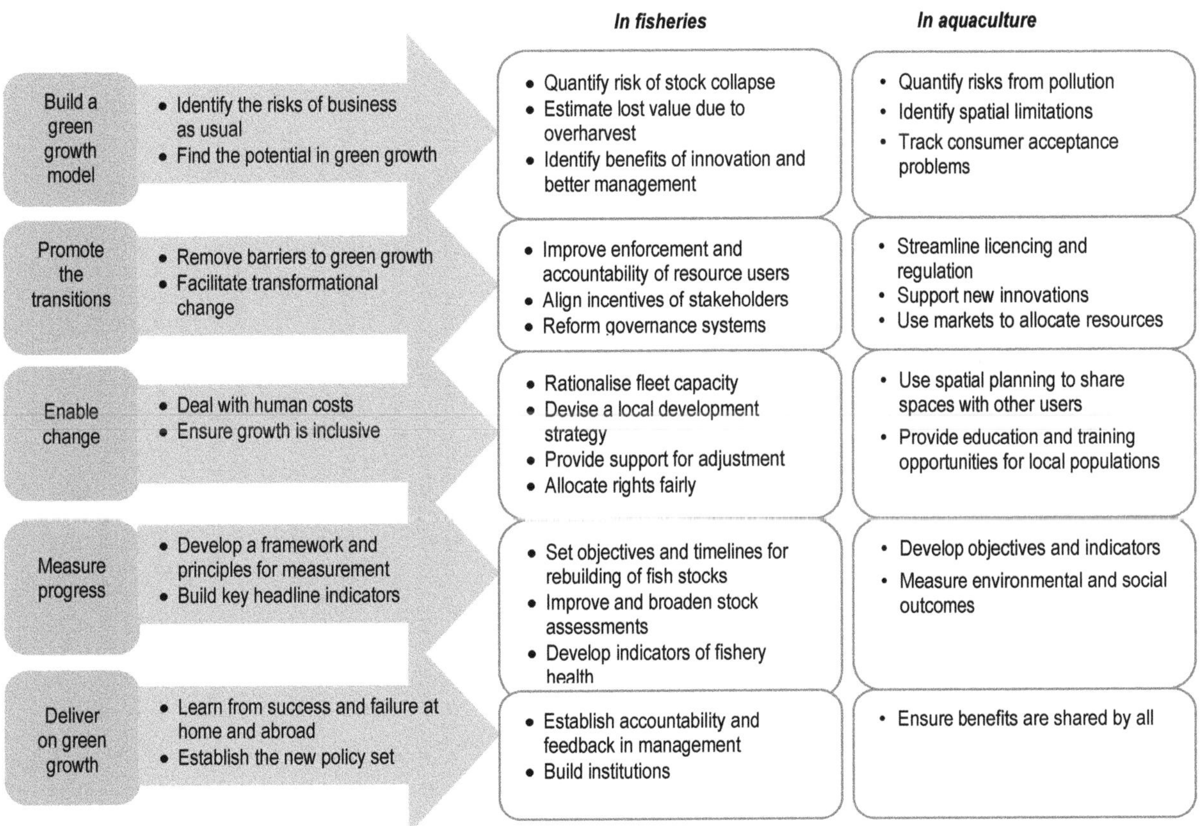

Source: Adapted from OECD (2010), "Interim Report of the Green Growth Strategy: Implementing our commitment for a sustainable future: Meeting of the OECD Council at Ministerial Level", 27-28 May, OECD internal document.

Table 1.1. Green growth policy toolkit: Some possible examples in fisheries

	In fisheries	*In aquaculture*
Regulations and standards	• Encourage use of product certification. • Strengthen standards for gear for selectivity and ecosystem preservation • Use area management including Marine Protected Areas where appropriate • Use landing obligation to reduce discards	• Simplified licencing procedures Zoning for use of costal and marine areas • Regulations concerning use of chemicals, medicines and additives
Support measures	• Help speed the adoption of environmental innovations in fishing gear • Target environmental outcomes where feasible, otherwise target actions and operations • Target public investments in green technologies	• Ensure access to capital for investment • Target environmental outcomes where feasible, otherwise target actions and operations • Target public investments in green technologies
Economic instruments	• Reform the use of fuel tax concessions to reflect external costs • Establish secure and tradable property rights in the fishery • Address policy constraints (governance, etc.) in less developed economies	• Emissions trading for pollutants from aquaculture • Establish secure tenure for spaces used for aquaculture
Trade measures	• Promote global markets and value chains • Eliminate export subsidies and restrictions • Support well-functioning markets	• Same as for fisheries
Research and development	• Increase public research on sustainable production systems • Promote private R&D through grants and tax credits • Undertake public/private partnerships for research	• Development of new species and production methods • Establishment of best practices
Development assistance	• Allocate more development aid for sustainable management initiatives • Raise profile of fisheries in Poverty Reduction Strategies • Allocate more funding for fisheries in Aid for Trade projects	• Provide services for smallholders, such as hatcheries and veterinary services • Extension services to promote adoption of best practices
Information, education, training and advice	• Increase public awareness for more sustainable patterns of consumption such as via eco-labelling and certification • Incorporate best practices in training, education and advice programmes throughout the entire value chain	Same as for fisheries

Source: Adapted from OECD (2013b), *Policy Instruments to Support Green Growth in Agriculture*, OECD Green Growth Studies, OECD Publishing, Paris.
DOI: http://dx.doi.org/10.1787/9789264203525-en.

Box 1.2. Green growth and sustainable development

Green growth has not been conceived as a replacement for sustainable development, but rather should be considered a subset of it. It is narrower in scope, entailing an operational policy agenda that can help achieve concrete, measurable progress at the interface between the economy and the environment. It provides a strong focus on fostering the necessary conditions for innovation, investment and competition that can give rise to new sources of economic growth – consistent with resilient ecosystems.

Green growth strategies need to pay specific attention to many of the social issues and equity concerns that can arise as a direct result of greening the economy – both at the national and international level. This is essential for successful implementation of green growth policies. Strategies should be implemented in parallel with initiatives centring on the broader social pillar of sustainable development.

Source: OECD (2011a), *Towards Green Growth*, OECD Green Growth Studies, OECD Publishing, Paris. DOI: http://dx.doi.org/10.1787/9789264111318-en.

One of the main ideas of the GGS is to bring together policy making in the economic and environmental domains, demonstrating the need for a bridge between these two policy areas that are often kept separate. Fisheries policy makers have been grappling with this problem for years when it became obvious that the seas were not a limitless resource and that fishing would need to be controlled if it was to be sustainable.

The green growth strategy has its roots in the ideas of sustainable development (Box 1.1). However, this strategy emphasises *policy action* and *deliverables* and so is a more tangible way of concurrently addressing economic growth and environmental concerns. With its focus on policy action and the need for long-term growth, this strategy is more operational for policy makers and provides more compelling arguments for policy reform. Green growth when applied to oceans is sometimes called "blue" (for example the EU *Blue Growth Strategy*, World Bank's *Global Partnership for Oceans*) (Box 1.2).

Governments have at their disposal a wide range of instruments for achieving green growth in fisheries and aquaculture (Table 1.1). In general, no one instrument or type of instrument can be singled out as more appropriate or efficient. The optimal mix of policy instrument depends on the objective to be achieved, and the environmental, economic, social and political context in which the instrument will operate.

Moving towards green growth

A number of policy barriers can hamper the move to green growth, including environmentally harmful subsidies, tariff and non-tariff barriers to trade in environmental goods and services, inefficient regulatory interventions and conflicting policy instruments (OECD 2010). Such barriers distort the efficient allocation of resources in the economy. In addition, when licences and other rights to fish lack flexibility such as the possibility to trade and consolidate them, fisheries resources will not generate the maximum benefits possible.

In the case of capture fisheries, reform towards a green growth path has been timid owing to a number of governance and institutional barriers to policy action. *Fisheries Policy Reform: National Experiences* (OECD, 2011d) *Rebuilding Fisheries: The way forward* (OECD, 2012) underscored that, while reform is necessary, resistance to change in fisheries is endemic. Particular factors inhibiting reform include i) the distribution of benefits and losses from change; ii) the timing of benefits and losses; iii) uncertainties regarding the distribution of benefits and losses; and iv) the ability to sustain reform processes over time. Reform strategies, including compensating those who lose out from reform, are therefore needed. In this process the role and involvement of stakeholder groups (e.g. fisher organisations, NGOs) is central to success and is a prerequisite for the alignment of incentive structures.

A green growth strategy must consider many specific details in designing policies for growth:

> *"One of the objectives of a green growth strategy is to find the policy mix that minimises the economic cost of a transition towards a growth path that better internalises environmental*

> *externalities. Therefore, one of the primary criteria for policy assessment is the cost-effectiveness of specific economic instruments. However, given the existence of monitoring and enforcement costs as well as information problems and market incompleteness, the appropriateness of policy instruments also needs to be assessed on the basis of their adoption and compliance incentives, of their ability to cope with uncertainty, as well as of their effectiveness in stimulating innovation and the diffusion of green technologies. Finally, given that environmental externalities often spill across national borders, the extent to which instruments can be designed and implemented in a way that facilitates international co-ordination is also considered."* (OECD 2010)

This description of green growth is especially relevant to the fisheries case as it observes the need to assess cost-effectiveness, underscores monitoring and enforcement costs as part of the equation, recognises market incompleteness, the role of incentives in adoption and compliance and the international dimension of the sector. These are longstanding and recognised issues in fisheries policy.

The OECD Green Growth strategy proposes a flexible policy framework that can be tailored to the different circumstances of each country (OECD 2011b). A range of options are available to promote positive change and establish a new policy set. The OECD suggests that policy makers build a new policy set that contains the following elements:

- Use market based instruments and pricing mechanisms when feasible.
- Ensure coherence across policy areas and instruments.
- Existing and new support programs should be carefully scrutinised for their side effects; environmentally harmful subsidies should be removed.
- Concurrently develop an innovation strategy which can help underpin a move to green technologies that addresses environmental externalities of production, distribution and consumption.
- Encourage consumers to demand sustainable products.
- Leverage private and public finance into green growth.
- Facilitate transition to green growth through labour market policies, improving skills and flexibility and adjustment to new economic sectors.
- In the face of adjustment income distribution may be a cause of concern for stakeholders and flanking policies may be needed to ensure buy-in. Measures to provide adjustment assistance or compensation can help enable reform.

Specifically for fisheries and aquaculture, to the above list can be added involvement of stakeholders, skills upgrading, flanking measures for the transition of workers, educational initiatives and innovation. In the case of capture fisheries focus should be on stock management and addressing overcapacity, whether in capital or labour. The traditional nature of fisheries means that specific solutions may look quite different in each country.

In the case of capture fisheries, *Using Market Mechanism to Manage Fisheries: Smoothing the Path* (OECD, 2006) outlines a list of ten tracks to follow to put in place the market-based approaches advocated in the GGS:

- Making all stakeholders comfortable with the concept of market-like instruments.
- Preferring an incremental or gradual implementation of market-like instruments.
- Not necessarily adopting a "one-size-fits-all" strategy.
- Carefully designing the process to allocate rights.
- Pragmatically using market forces.

- Overcoming the "excessive consolidation" question.
- Using the "demonstration effect" (drawing on experience).
- Involving stakeholders in the reform process.
- Integrating fisheries characteristics.
- Dealing pragmatically with trade-offs.

The world is increasingly globalised and interconnected. This is particularly the case in fisheries and aquaculture where more than 50% of all fish is traded across the globe and a substantial amount of the fish consumed in OECD countries originates in developing countries. Many fish stocks are also global in the sense that they are crossing national boundaries because they are migratory (e.g. tuna) or the stocks may seasonally move in and out or between national fishing zones. It follows that there is a strong need for international co-operation if we are to achieve green growth in fisheries. *Fishing for Coherence* (OECD, 2006) and *Strengthening Regional Fisheries Management Organisations* (OECD, 2009) show that strengthening international cooperation in fisheries management, in particular for development, can bring real benefits to fishers and for sustainability.

The OECD has developed a conceptual framework for monitoring progress towards green growth, including a set of indicators (OECD, 2011c). This work observes that "green growth indicators should … identify the most important environmental services, and if possible quantify their role in economic growth". While the set of indicators is still being refined, the headline indicator pertinent to the fisheries and aquaculture sectors measures

> *"The proportion of fish stocks within safe biological limits (global), expressed as the percentage of fish stocks exploited within their level of maximum biological productivity, i.e. stocks that are underexploited, moderately exploited, and fully exploited. Safe biological limits are the precautionary thresholds advocated by the International Council for the Exploration of the Sea (ICES). This indicator is also included in the Millennium Development Goal monitoring framework. Trends in fish production from aquaculture along with trends in fish production from capture fisheries presented worldwide and for major species groups are given as complements".*

Making progress in fisheries

Over the past decades several countries have introduced market-based instruments in their fisheries management. In *Fisheries Policy Reform: National Experiences* (OECD, 2011d), it was noted that "*poor environmental performance is generally not sufficient to prompt governments to undertake significant reform efforts. In each of the case studies, it was economic crisis rather than environmental crisis that provided the key trigger for reforms to be contemplated and acted upon*". A consequence of this is that the overall costs induced by poor environmental performance – for private operators and for public coffers as forgone revenue – will be higher the longer one waits to undertake reform. This is in line with the observations that delaying reform will result in continued rent dissipation, higher risk for fish stocks, and generally lower fish stocks than otherwise would have been the case with reform and rebuilding.

Fisheries management reform can bring considerable economic and social benefits. The Danish experience has been that better green growth strategies in fisheries increase stock sizes, reduce the amount of inputs used (vessels, fishers, energy, etc.), improve catch per unit of effort (CPUE) and could also increase quality of harvest (Box 1.3). This makes efforts to overcome resistance to reform worthwhile. As noted in *Towards Green Growth "Consulting on how distributional impacts will be dealt with is a crucial part of policy communication. This includes taking careful account of how affected groups want to be compensated".*

Box 1.3. Market-based instruments in Denmark

In Denmark, a system of individual transferable quotas was introduced in 2003 in the pelagic fishery. In 2007, the scheme (also somewhat modified) was introduced in the remaining fisheries. The consequence has been that the number of active vessels declined from 1 528 in 2000 to 688 by 2010. Simultaneously, the total operating profits increased from DKK 620 million in 2006 to DKK 1453 million in 2010, corresponding to a 130% increase. The Danish example shows that green growth can be achieved through better fisheries regulation and fisheries policy reform. Other green growth improvements have included a considerable reduction in energy use.

	2006	2007	2008	2009	2010
Number of registered vessels[1]	3 134	2 957	2 890	2 834	2 826
Number of commercially active vessels[1]	1 093	846	777	703	688
Number of employed	2 341	1 751	1 577	1 446	1 392
Total landing value (DKK Million)	3 183	2 719	2 560	2 218	3 004
Average per commercially active vessel					
Landing value (DKK 1 000)	2 785	3 053	3 076	2 955	4 176
Earning (DKK 1 000)	1 726	1 857	1 691	1 636	2 658
Operating profit (DKK 1 000)	620	829	609	623	1 453
Net profit (% of insurance value)	15%	20%	20%	13%	30%

Source: OECD (2013), *OECD Review of Fisheries: Policies and Summary Statistics 2013*, OECD Publishing, Paris. DOI: http://dx.doi.org/10.1787/rev_fish-2013-en.

One way to ensure that reforms bring net benefits – to both private operators and public coffers – is to rely more strongly on cost recovery to fund fisheries management (Box 1.4). A cost recovery approach ensures that the management system is delivering net benefits to fishers and so acts as a benchmark for efficient management. For example in New Zealand the government, through a series of levies on quota holders, catch limits conservation service levies recoup around one-third of the total costs of "general services" to the fishing industry. Cost-recovery approaches will always be country- and fisheries management-specific (Table 1.2). In some cases (Iceland, New Zealand) the right institutional frameworks have led the industry to demand additional expenditures on management services suggesting that well-designed reforms can demonstrate the benefits of sound management.

Some reforms can bring lower costs, and policy reforms are a good moment to consider ways to reduce management costs. There is some evidence to suggest that *"Those OECD countries employing predominantly output controls have significantly lower total costs per tonne of production than the rest of the OECD countries. They also mostly have lower management and enforcement costs per tonne of production*" (OECD 2003).

Box 1.4. Cost recovery in fisheries

The question of how to improve and make the delivery of management services more cost effective is an important one for policy makers. Fisheries management costs are an important factor in delivering effective and sustainable fisheries management outcomes. Budgetary pressures mean that obtaining the best value for money and resource efficiencies are always a priority for policy makers.

Using cost recovery to fund fisheries management shifts the burden from the taxpayer to the beneficiary of the management services. If the benefits of fisheries management are less than their costs, cost recovery will make this apparent and serve as a prod to reform that either reduces costs or changes the economics of the fishery.

Institution structure determines how users are co-ordinated, how information is generated, how decision are made and how monitoring and enforcement take place and will influence the magnitude of the costs of arranging, monitoring and enforcing agreements. The extent of user participation in the process of design, implementation and enforcement of fisheries management regimes will also influence services costs.

Source: OECD (2003), *The Costs of Managing Fisheries*, OECD Publishing, Paris. DOI: http://dx.doi.org/10.1787/9789264099777-en.

Table 1.2. Typology of fishery service delivery and payment models

Model	What type and level of services to provide	Who provides the services	Who pays for the services	Incentives for improved fishery performance	Incentives for increased cost-effectiveness
1	Government	Government	Government	Poor	Poor
2	Government	Government	Industry	Poor	Poor
3	Government	Government contracts service providers	Industry	Poor	Good
4	Government and industry	Government contracts service providers	Industry	Good	Good
5	Government sets standards and industry undertakes decisions	Devolved to industry	Industry	High	High

Source: OECD (2003), *The Costs of Managing Fisheries*, OECD Publishing, Paris.
DOI: http://dx.doi.org/10.1787/9789264099777-en.

As management becomes more holistic, data and information requirements become increasingly complex and costly. Achieving higher levels of coherence across policy domains affecting the broader use of oceans and requires high levels of institutional co-operation (Charles, 2011). This co-operation can be time consuming, so finding ways to streamline cooperation will bring additional benefits.

Finding new Green Market Opportunities

The OECD GGS emphasises that new green market opportunities can be a source of growth as economies transition to a greener growth path. In particular, investing in innovation and taking the "first mover" advantage are two ways to ensure that governments can take advantage of new sources of economic growth.

Recreational fishing can in some cases deliver more economic value per quantity harvested than commercial fishing. In 2009, commercial fisheries in the United States induced a turnover of USD 116 billion compared to USD 50 billion for the recreational sector (Box 1.5). In the same year jobs created or sustained were 1 million full/part time in commercial fishing compared to 327 000 jobs supporting recreational fishing excluding the 11 million recreational anglers taking part in this activity.

Policy tools to manage different economic activities sharing a common resource will in principle include use of auctions and other means of making use of the resource transferable between activities. In practice, most countries use command and control methods to delineate the activities of commercial vs. recreational fishers and others. This may include specific licensing arrangement for recreational fisheries, prohibition on selling the fish (thus avoiding market interaction and the potential under-pricing of commercial fish), limitations on gear and seasons. The reason for this may have to do with the transactions costs or other limitations that make exchange of use rights difficult. A green growth strategy would look for solutions to reduce or abolish these limitations in order to maximise the benefits derived from the resource.

Commercial and recreational fisheries often exploit the same fish stock and can be in competition with each other. Both activities generate economic value albeit in different ways. To ensure that the fish stock is exploited in the best way possible, the fishery manager must set the optimal joint total allowable catch (TAC) for both fisheries and allocate this TAC optimally to both sectors. As highlighted by Arnason (2012), "*commercial fisheries are operated primarily for profits and recreational fisheries primarily for enjoyment. There are important similarities, however. Both extract from the fish stocks and thus have a comparable impact on the evolution of fish stocks. In particular, recreational fisheries will, just as commercial fisheries, reduce the size of fish stocks although the details of the impact may be different. Moreover, recreational fisheries are no less than commercial fisheries subject to the ill-effects of lacking property rights and, consequently, excessive effort and overexploitation".* Arnason concludes that the economic importance of the recreational fisheries sector should be taken into account in allocations and that those allocations would be most efficiently achieved through a market based solution such as a tradable rights system.

Box 1.5. Recreational vs. commercial fishing in the United States

The US seafood industry includes the commercial harvest sector, seafood processors and dealers, seafood wholesalers and distributors, importers, and seafood retailers. In 2009, this industry supported approximately 1 million full- and part-time jobs and generated USD 116 billion in sales impacts, USD 32 billion in income impacts, and USD 48 billion in value added impacts.

Seafood retailers generated the highest job and income impacts, contributing 484 000 jobs and USD 10 billion in 2009. In contrast, the largest sales (USD 49 billion) and value added impacts (USD 15 billion) came from the importer sector. The seafood wholesalers and distributors sector contributed the least to the national seafood industry impacts with 47 000 employees, USD 6.5 billion in sales impacts, USD 2.1 billion in income impacts, and USD 3.1 billion in value added impacts.

In 2009, there were approximately 11 million recreational anglers across the United States who took 74 million saltwater fishing trips around the country. These anglers spent USD 4.5 billion on fishing trips and USD 15 billion on durable fishing-related equipment. These expenditures contributed USD 50 billion in sales impacts to the US economy, generated USD 23 billion in value added impacts, and supported over 327 000 job impacts.

Durable equipment impacts contributed the most to these totals, accounting for 74% of employment impacts, 79% of total sales impacts, and 77% of value added impacts. Of the three fishing trip modes, shore-based fishing trips contributed the most to the number of jobs supported by recreational angling with 11% of employment impacts. For-hire sales (USD 1.9 billion) and value added impacts (USD 1 billion) were approximately half the magnitude of impacts generated by either private boat (USD 4.2 billion, USD 2.2 billion) or shore-based trips (USD 4.3 billion, USD 2.2 billion).

Source: National Marine Fisheries Service (2010), *Fisheries of the United States 2009*, Silver Spring, Maryland.

Innovation and improved technologies

The report Towards Green Growth (OECD, 2011a) suggests that "The core of transforming an economy is innovation. Innovation and the resulting creative destruction mean new ideas, new entrepreneurs and new business models. It contributes to the establishment of new markets, leads to the creation of new jobs and is a key ingredient of any effort to improve people's quality of life." Also, innovation is fundamental to addressing environmental issues (Box 1.6). Achieving green growth in practice means increasing economic output while maintaining or reducing demand on environmental inputs. When fisheries management and governance systems are doing their job well, the only way to obtain continued growth is through innovations that add value, create new markets, or lower costs. In fisheries, for example, trawl door design that do not rip up the sea floor is a major advance in leaving the sea floor ecosystem undisturbed and can produce wider social benefits.

New techniques and innovations can spur growth through improving quality and catch quantity, by reducing costs of harvesting or by creating increased private and social benefits. Some innovations will be entirely in the private domain and be spurred by the search for additional profits. Other innovations are driven by the public sector advances in technology, costs reductions (e.g. surveillance) and enforcement and through investments by the public sector.

Box 1.6. Innovation for Green Growth

Innovation will be an important driver of the transition towards green growth. Without innovation, it will be very difficult and very costly to achieve a transformation to a greener economy. By pushing the frontier outward, innovation can help to decouple growth from natural capital depletion. Innovation and the related process of creative destruction will also lead to new ideas, new entrepreneurs and new business models, thus contributing to the establishment of new markets and eventually to the creation of new jobs. Innovation is therefore the key in enabling green and growth to go hand-in-hand

Source: OECD (2011e), *Fostering Innovation for Green Growth*, OECD Green Growth Studies, OECD Publishing, Paris. DOI: http://dx.doi.org/10.1787/9789264119925-en.

The OECD Innovation Strategy identifies five principles that can help governments use innovation (including green innovation) as a tool to improve economic performance, address societal challenges and enhance welfare. The five principles are i) empowering people to innovate; ii) unleashing innovations; iii) creating and applying knowledge; iv) applying innovation to address global and social challenges; and v) improving the governance and measurement of policies for innovation. A green growth strategy in fisheries would find practical approaches to applying these principles to the fisheries sector.

While the private sector is the main driver of innovation public policies can support a move towards green innovation. OECD work suggests that in cases where market failures stymie green innovation, public intervention to address those market failures may be warranted (OECD, 2011e). Boosting green innovation requires clear and stable market signals and public investment may be required for basic and long term research, possibly through international cooperation. Finally the work underlines that specific public interventions may be necessary to support private investment in innovation, R&D, support for general purpose technologies, fostering new entrepreneurial firms and through facilitating transition to green technologies in small and medium sized enterprises.

In fisheries and aquaculture, innovations have been driven to a large extent by evolving legislative frameworks for the sustainable use of fish and other natural resources. For example, size limits on fish, discards and by-catch rules, requirements for fishing gear will induce net-makers and fishers to search for new solutions and improved fishing techniques. Another driver has been the search for lowering cost of energy use in fishing and includes hull and gear design and in particular motor technology and the installation of effective fuel meters.

Innovations to reduce waste have led to new uses for fish parts in glue, extraction of bioactive peptides (from fish bones and frames, heads), chitin (from shrimp shells), fish sauces, fish skin for leather goods and more. The basic point is that when a market exists or can be created/developed (e.g. fish oil supplements) innovation and products will follow. The introduction of a discard ban in the European Union, starting in 2014, will lead to the landing of fish without a ready consumer market. Alternative uses and markets – e.g. ensilage, fodder – will need to be developed to bring the raw material to good use.

An example of innovation in the public domain is the use of remote electronic monitoring through a combination of CCTV (closed circuit television) technologies and electronic log book data (Box 1.7). This can reduce monitoring costs by eliminating observers while at the same time increasing coverage and effectiveness. Denmark and the United Kingdom have been very active in this field making it easier to monitor and control fishing activities. A so-called "fully documented fishery" entails detailed recordings in logbooks combined with the use of electronic monitoring systems where various sensors and CCTV cameras are recording fishing events and catch handling operations. Sensor recordings and video footage make it possible to retrospectively verify the electronic logbook records.

Box 1.7. Danish experiments with monitoring and remote sensing

Catch Quota Management (CQM) including full documentation has been on trial in Danish fisheries in the period 2010 to 2012. By introducing full accountability through catch quotas instead of landing quotas the fisherman's incentive to optimise the value of his catch by discarding less valuable fish would be substituted by his incentive to use selective fishing methods to optimise the value of his total removals from the stocks. The trial aimed at testing whether CQM could provide a reliable accounting for all catches of cod, give better scientific data and encourage fishermen to fish more selectively and reduce accidental catches. The main feature of the trial is that all catches count against the vessel quota and that the fishing vessels are monitored from port to port using sensors and CCTV technology. The Remote Electronic Monitoring (REM) system has collected sensor data and images throughout the trial period and according to the vessel electronic-logbooks the vessels were at sea for approximately 80 000 hours, carried out approximately 1 114 fishing trips and conducted approximately 9 800 fishing operations during the project period.

Source: Ministry of Food, Agriculture and Fisheries of Denmark, *Yield of Fish* accessed via http://agrifish.dk/fisheries/fishery-statistics/

Box 1.8. The use of fuel flow meters in the United Kingdom

Fuel flow meters proved useful to skippers in highlighting the change in fuel consumption as a result of altering engine revs. Small reductions in revs resulted in substantial reductions in fuel usage. This information proved very useful in setting the optimum revs of the engine whilst the vessel was steaming to the fishing grounds with a minimal loss to vessel speed. Several of the skippers commented that this information alone would result in the meter paying for itself in less than six months.

Source: Seafish (2008), "Work currently being undertaken by Seafish to improve fuel efficiency in the fishing fleet", *Factsheet/Datasheet*, April.

The high cost of energy faced by many fishers has driven innovation in energy efficiency. The UK's Seafish is an example of how the public and private sectors can work together to promote innovation. Seafish is funded by a levy on the first sale of seafood landed and imported in the United Kingdom. It aims to support and improve the environmental sustainability, efficiency and cost-effectiveness of the industry, as well as promoting sustainably-sourced seafood. Seafish has worked on the installation of fuel meters on fishing vessels as well as other energy saving devices notably gear and vessel design. This type of research has proven valuable for the UK fishing industry (Box 1.8).

The Nordic Marine Innovation (part of the Nordic Council) has funded a number of marine related innovation programmes and innovation conferences. This includes a project related to the removal of pinbones in cod and whitefish, the development of novel bioactive seaweed-based products, marketing, including retail displays of fish and fish products. A particularly interesting project is the establishment of a North Atlantic Marine Cluster which seeks to strengthen relationships, build arenas for communicating research findings and benchmarking in the marine sector.

Improving the efficiency of resource use has a central role in green growth planning. In fisheries, there remains significant room for improvement in the use of discards, by-catch, and trimmings, among other things. One part of the OECD green growth strategy is to reform policies that impede progress, such as requirements that fish must be landed headed and gutted, or that fish for which the fisher does not have a quota must be discarded. A sector wide industrial dialogue may be a useful tool to identify where in the value chain improvements can be made.

Other elements of the OECD green growth strategy are to identify opportunities for green growth and to put in place a policy set that helps realise these opportunities. For example, recent developments show that by-products can have higher value if used in functional foods and in pharmaceuticals. For example, bioactive peptides have shown to be good inputs into promoting health as they reduce blood pressure, can address cancer, obesity and diabetes. In addition, fish skins can also be used in the leather garments industry. The report "*Marine Biotechnology: Enabling Solutions for Ocean Productivity and Sustainability*" (OECD, 2012b) considers *inter alia* the potential of marine biotechnologies and the challenges of developing marine bio resources as these are situated in complex marine ecosystems where our knowledge is currently limited. A key conclusion is the need for international co-operation in developing these resources and the building of a monitoring framework – including economic indicators – which will help understand the potential direction of this marine green growth potential.

Several new species have come into aquaculture production over recent years (e.g. *pangasius*) and scientists are working on bringing the life cycle of other species under control. The principal challenge is feeding at the early stages of larvae after hatching. In addition from one species to the other only little of the knowledge and techniques are transferable. In other words bringing a new species under human control requires major laboratory efforts and starting fresh for each species. Species such as trout and salmon are mature in production, while some species with great potential such as the Bluefin tuna and the eel have not yet seen full control of the reproduction circle.

The *pangasius* story is one of remarkable success. Production in Vietnam has grown from around 100 000 tonnes in 2000 to more than 1.4 million tonnes in 2008, but such rapid growth has led to environmental and social concerns. Another fast growing production has been shrimps, facilitated by the development of pathogen-free and pathogen-resistant species. A ready consumer market, logistics and processing capabilities are of course central to these developments.

The reduction in the use of fishmeal and in particular fish oil from feed compounds for carnivorous species remains a particular concern. Sustainable growth of aquaculture production requires sources of feed that can also grow, and many wild species currently used are unlikely to be able to support increased harvest levels. Research into using various new sources of raw material as feed for fish farming is seen as a solution. These include the use of algae and mussels (Nordic Innovation initiative) in trout and tilapia farming and changing feed compounds for turbot aquaculture. Such substitution of novel feed sources can also provide significant costs savings.

Alternative systems of aquaculture production are being developed with a view to reducing environmental load. Recirculating systems and integrated multi-trophic aquaculture productions systems are among the ways proposed to capturing the nutrients that would otherwise be discharged to the environmental (Box 1.9).

Box 1.9. Canadian Aquaculture Innovation and Market Access Programme (2008-2013)

The overall goal of the Aquaculture Innovation and Market Access Program (AIMAP) was to catalyse aquaculture industry investment from the private sector, as well as other sectors, that would:

- Improve the competitiveness of the Canadian aquaculture industry by encouraging an aquaculture sector that continuously develops and adopts innovative technologies and management techniques to enhance its global competitiveness and environmental performance; and
- Position Canadian aquaculture products as having high value in the market place based on their environmental performance, traceability and other considerations.

These goals contributed to the DFO strategic outcomes of sustainable fisheries and aquaculture and healthy and productive aquatic ecosystems. Contribution funding under AIMAP was intended to enable recipients to plan, manage and complete projects that would achieve these strategic outcomes. The AIMAP was a nationally competitive process with calls for proposals issued on an annual basis, and based on priorities established in consultation with provinces, territories and sector stakeholders. The program focused on one year duration projects with demonstrable industry-wide benefits implemented by the end of the project.

Under this program a maximum of USD 4.5 million in DFO contribution funds was available annually to support innovation initiatives, and up to USD 0.2 million in DFO contribution funds was to support market access initiatives.

Source: Fisheries and Oceans Canada (2014), *Aquaculture Innovation and Market Access Program (2008-2013)*, http://www.dfo-mpo.gc.ca/aquaculture/sustainable-durable/index-eng.htm.

Green growth indicators for fisheries and aquaculture

The OECD's green growth indicators set is currently proposing only one green growth indicator for fisheries: the proportion of fish stocks within safe biological limits (global), expressed as the percentage of fish stocks exploited within their level of maximum biological productivity. While this is an important indicator, it does not capture the full range of issues fisheries policy makers will need to monitor as they implement a green growth agenda.

A good indicator should be unambiguous - an increase in its numerical value should be either always good or always bad. Total landings are not a good indicator because increases are good if they are done sustainably and bad otherwise; it needs more information to be understood. The same is true for other economic measures like employment, number of vessels, or profits, where understanding whether growth is "green" or not is important. On the other hand, quota rent is a good measure as this ideally represents the net present value of the fishery, and so incorporates sustainability implicitly.

Headline indicators are effective ways to summarise the situation for a fishery to the public. Monitoring progress and measuring results is an essential part of the OECD GGS, but can be a complicated business (Box 1.10). Summary indicators can present key messages in an approachable way and can help galvanise public support for change. For example, "50% more food is required by 2050" encapsulates a message regarding the pressures faced by the food system due to population growth where the underlying issues are actually quite complex. The often-quoted "USD 50 billion per year in additional revenue if fisheries are optimally managed" (World Bank 2009) is probably over-precise, but gives the message that the benefits are large and achievable.

Box 1.10. Headline indicators

The multi-dimensional nature of green growth requires a sufficient number of indicators to do justice to the various aspects of the issue at hand. But a large dashboard also carries the danger of losing a clear message that speaks to policy makers and helps communication with the media and with citizens. One way of addressing this issue is to construct a composite indicator. The advantages of ease of communication and concise presentation of a composite number must, however be weighed against the problem of choosing units and weights required for aggregation across very different elements. ...it is proposed that a small set of "headline" indicators be selected that are able to track central elements of the green growth concept and that are representative of a broader set of green growth issues. This is a task that still lies ahead and requires broad consultation and discussion because, inevitably, opinions on the most salient set of indicators will vary among stakeholders. The OECD stands ready to take this task forward.

Source: OECD (2011a), *Towards Green Growth*, OECD Green Growth Studies, OECD Publishing, Paris.
DOI: http://dx.doi.org/10.1787/9789264111318-en.

The OECD can help with the provision of headline indicators. The Organisation is not only working on defining and producing good indicators as part of its overall green growth work, but the data collected by the Committee for Fisheries can help form the basis of such indicators. The Government Financial Transfers (GFT) database measures and classifies support provided to fisheries, an important indicator of either government commitment to the sector or the potential for reform.

More generally, good indicators should be able to monitor trends and structural changes, attract attention to issues that require further analysis and possible policy action and help measure how well policies are performing with respect to green growth (OECD, 2011b). The key principles suggested by the OECD may help the selection of relevant green growth indicators (Box 1.11). Indirect indicators can be useful when direct ones are impractical or ambiguous. For example, measuring the availability or use of energy audits demonstrates increases in environmental awareness on the part of fishers and indirectly measures energy efficiency.

Based on these key principles several green growth indicators are proposed for a number of aspects of fisheries and aquaculture production (Table 1.3). However, the proposed indicators do not pretend to be exhaustive or final.

Box 1.11. Key principles in selecting indicators to monitor progress in green growth

Policy relevance	The indicator set should have a clear policy relevance and in particular: • Provide a balanced coverage of the key features of green growth with a focus on those that are of common interest to OECD member and partner countries. • Be easy to interpret and transparent. • Provide basis for comparisons across countries. • Lend itself to being adapted to different national contexts and analysed at different levels of detail or aggregation.
Analytical soundness	The indicators should be analytically sound and benefit from a consensus about their validity. Th further lend themselves to being linked to economic and environmental modelling and forecasting.
Measurability	The indicators should be based on data that are available or that can be made available at a reason and that are of known quality and regularly updated.

Note: These principles and criteria describe the "ideal" indicator; not all of them will be met in practice.

Source: OECD (2011b), *Towards Green Growth: Monitoring Progress: OECD Indicators*, OECD Green Growth Studies, OECD Publishing, Paris.
DOI: http://dx.doi.org/10.1787/9789264111356-en.

Table 1.3. Some potential green growth indicators in fisheries

Indicator	What it measures	Notes
Fisheries policy indicators		
GFT as % of value	Measures scale of policy effort relative to scale of sector. Can indicate degree of dependency of sector on support; social value versus economic importance of sector	The denominator of this indicator scales the level of GFT; it can be value of landings, number of vessels or similar. Different denominators focus on different aspects of the sector.
GFT Composition	Measures focus of policy effort and scope of objectives addressed by policies. Can indicate impacts of policies in certain cases	This indicator can form the basis for analysis to measure impacts of policies but probably only has a limited capacity to do so on its own.
Cost-sharing of fisheries management	Measures share of management costs paid by fishers. Can indicate net value of sector.	Related to the %GFT measure in that it indicates scale of policy support, but with a focus on management costs.
Biological indicators		
Stock Status	Primary indicator of fishery health and sustainability	Stock size can be measured as total or spawning biomass, measured relative to MSY or MEY levels
Biodiversity	Measure of resilience and productivity of ecosystem	Notwithstanding the importance placed on managing biodiversity, the relationship between fisheries, aquaculture and biodiversity is complicated and indicators are difficult to construct.
Economic indicators for fisheries		
Value per vessel	Measures economic concentration and revenue; it is a weak indicator of profitability and income.	This is a proportional indicator; there are several similar ones that could be created with value or catch quantity in the numerator and vessels, companies or fishers in the denominator
Fleet capacity	Measures level of capitalisation in the fishery. Indicates potential for overcapacity	A related proportional indicator would be the ratio of active vessels to all vessels. Technical overcapacity is not always bad, complicating the interpretation of this indicator.
Quota value	Measures rents generated by fishery	This may be a more useful measure of capacity and profitability, but is only available where individual quotas are in place.
Fisheries management indicators		
Presence of a harvest control rule	Measures technical capacity in stock management	Harvest control rules are systematic approaches to stock management that define allowed harvest levels for a given (measured) stock size
Presence of appeals mechanism	Measures quality of governance	This is one of the criteria for good governance outlined in this report
Presence of formal venues for stakeholder participation	Measures quality of governance	This is one of the criteria for good governance outlined in this report

Table 1.3. Some potential green growth indicators in fisheries

	Fisheries management indicators (*cont.*)	
Share of landings that are IUU	Measures quality of enforcement	Can be difficult to estimate; alternatives include MCS expenditures, compliance with international agreements, flag state rules
Presence of overfishing	Measures quality of TAC setting and enforcement	Related to stock status—a fishery can be either "overfished", "subject to overfishing" or both. This can be complicated by recovery plans that allow temporary overfishing
	Aquaculture indicators	
Feed conversion ratio (FCR)	Resource use efficiency (amount of feed ingredients such as fishmeal and fish oil, feed composition, feeding technology, feed loss, fish growth and disease), productivity, amount of discharge	Fed farming systems
Water quality	Environmental impact on surrounding waste, disease risk	Fed farming systems
Escapees	Biological risk	Fish farming systems except RAS system
Medicine	Disease risk, human health risk	All farming systems
Energy use	Contribution to climate change, Resource use efficiency	All farming systems
Water use	Resource use efficiency	Freshwater farming systems
Land use	Resource use efficiency	Inland farming systems
GDP contribution	Economic impact	All farming systems
Job creation	Economic impact	All farming systems

In Norway, work is ongoing on establishing a set of measurable indicators of green growth in the salmon farming sector. The most important environmental challenges to solve for the Norwegian aquaculture industry in the short-term are escape of farmed salmon into the wild and the spreading of sea lice from aquaculture sites. The Norwegian government has initiated work to develop and implement a concept of *indicators of sustainability*, together with action limits, to manage these two environmental challenges (Figure 1.5). The concept is based on a proposal from the Norwegian Institute of Marine Research and the Norwegian Veterinary Institute, with contribution from the Norwegian Institute for Nature Research. The government will stimulate further research and will evaluate and continuously update both indicators and action limits taking into account state of the art knowledge.

Going for green growth

Aquaculture is expected to be the main source of growth in supply of fish products in the future. The growing role of aquaculture does not need to come at the expense of capture fisheries, but nevertheless it foreshadows a changing world to which capture fisheries and fisheries managers will need to respond.

Figure 1.5. A concept of sustainability indicators in Norway

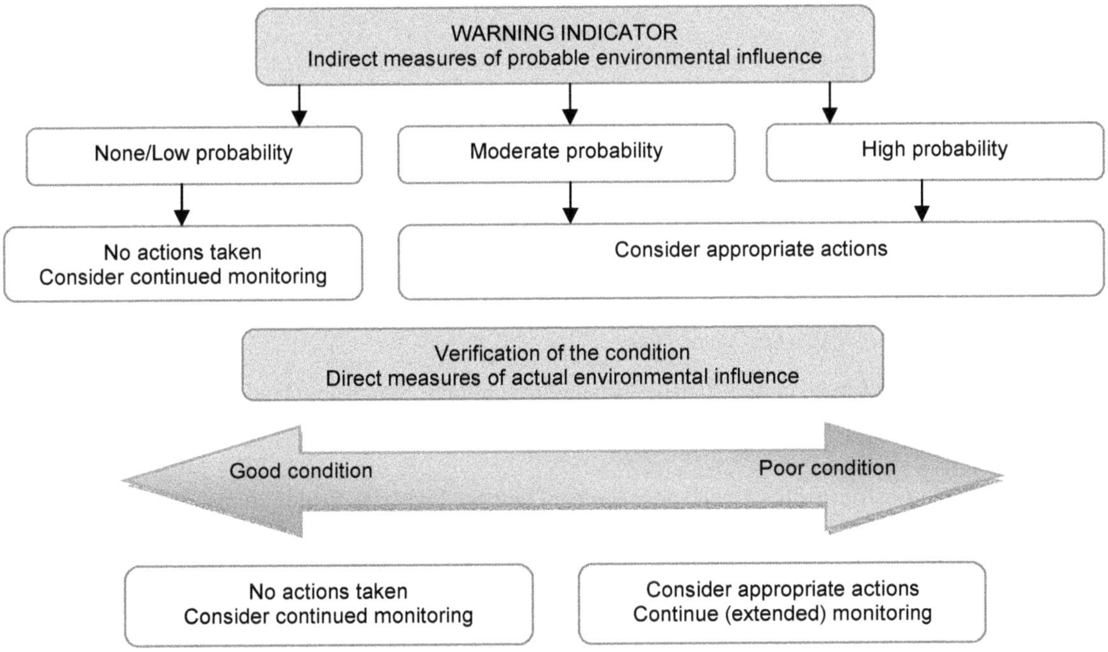

Source: Norwegian Ministry of Fisheries and Coastal Affairs.

Green growth reforms can help ensure that capture fisheries are competitive now and in the future. Markets are increasingly global and aquaculture will be a growing and an effective competitor for an increasing number of species. Other users of marine resources will also increase their productivity and scale of operations and become more competitive users of marine inputs. Meeting the competitiveness challenge requires a focus on economic efficiency. For policy makers intent on securing the future of capture fisheries alternative regulatory frameworks that allow capture fishers to operate in the best, most productive way need to be implemented to be able to compete with aquaculture.

In both fisheries and aquaculture many innovative approaches have helped underpin growth. Innovation will continue to be a strong driver of growth and new technologies have the potential to address the conflicts and limitations arising from the interactions between capture fisheries and aquaculture. There is a role for public intervention to promote innovation, both through government research and promotion of private research and development.

The green growth concept has underscored that a "business as usual" policy in capture fisheries management inevitably, in many cases, will lead to low fisher incomes, excessive costs and lost revenue for national government coffers. In times with increasing focus on government expenditure the fisheries policy makers can make a major contribution by spearheading fisheries reform. Green growth reform in capture fisheries will improve the environment, contribute to a better fisheries economy and increase the tax base bringing fisheries back from a potentially marginalised position. Such a win-win-win approach may meet resistance among certain groups and bringing all concerned stakeholders along is a prerequisite for successful reform. Policy packages underpinning reform which addressees concerns by those groups that stand to lose will be necessary including effective flanking measures. But overall society stands to gain.

References

Asche, F. (2008), "Farming the Sea", *Marine Resource Economics*, Vol. 23, pp. 527–547.

Arnason, R. (2012), *Managing Commercial and Recreational Fisheries: Issues and challenges*, (OECD internal report).

Bell, J.D., K.M. Leber, H.L. Blankenship, N.R. Loneragan, and R. Masuda (2008), "A new era for restocking, stock enhancement and sea ranching of coastal fisheries resources", *Reviews in Fisheries Science*, Vol. 16(1-3), pp. 1-9.

Bostock, J., B. McAndrew, et al. (2010), "Aquaculture: Global Status and Trends", *Philosophical transactions of the Royal Society,* Vol. 365: 2897-2912
DOI: 10.1098/rstb.2010.0170.

Charles, A. (2011), "Integrated Ocean Management and the Fisheries Sector", Consultant Report (OECD internal document).

FAO (2012), *State of the world's fisheries*, Food and Agriculture Organization Publications, Rome.

Garcia, S.M. and A.A. Rosenberg (2010), "Food Security and Marine Capture Fisheries: Characteristics, Trends, Drivers and Future Perspectives", *Philosophical Transactions of the Royal Society,* Vol. 365: 2869-2880 .
DOI: 10.1098/rstb.2010.0171.

IFFO (2012), "Aquaculture and Green Growth workshop", Presentation given at the Green Growth and Aquaculture Workshop, 2-13 December 2012, Yeosu, Korea, Consultant Report (OECD internal document).

Jackson, J.S. (2010), "Connections between farmed and wild fish: Fishmeal and fish oil as feed ingredients in sustainable aquaculture", in OECD, *Advancing the Aquaculture Agenda: Workshop Proceedings*, OECD Publishing, Paris.
DOI: http://dx.doi.org/10.1787/9789264088726-22-en.

National Marine Fisheries Service (2010), *Fisheries of the United States 2009*, Silver Spring, Maryland.

OECD/FAO (2014), *OECD-FAO Agricultural Outlook 2014*, OECD Publishing, Paris.
DOI: http://dx.doi.org/10.1787/agr_outlook-2014-en.

OECD (2013a), *OECD Review of Fisheries: Policies and Summary Statistics 2013*, OECD Publishing, Paris.
DOI: http://dx.doi.org/10.1787/rev_fish-2013-en.

OECD (2013b), *Policy Instruments to Support Green Growth in Agriculture*, OECD Green Growth Studies, OECD Publishing, Paris.
DOI: http://dx.doi.org/10.1787/9789264203525-en

OECD (2012a), "Summary of Discussions/Presentations and Case Studies", Green Growth and Aquaculture Workshop, 2-13 December 2012, Yeosu, Korea, consultant report, OECD internal document.

OECD (2012b), "Marine Biotechnology: Enabling Solutions for Ocean Productivity and Sustainability", consultant report, OECD internal document.

OECD (2011a), *Towards Green Growth*, OECD Green Growth Studies, OECD Publishing, Paris.
DOI: http://dx.doi.org/10.1787/9789264111318-en.

OECD (2011b), *Towards Green Growth: Monitoring Progress: OECD Indicators*, OECD Green Growth Studies, OECD Publishing, Paris.
DOI: http://dx.doi.org/10.1787/9789264111356-en.

OECD (2011c), *Fisheries Policy Reform: National Experiences*, OECD Publishing, Paris.
DOI: http://dx.doi.org/10.1787/9789264096813-en.

OECD (2011d), "Tools for Delivering on Green Growth", Paper prepared for the 2011 OECD Ministerial Council Meeting 2011, Paris.

OECD (2011e), *Fostering Innovation for Green Growth*, OECD Green Growth Studies, OECD Publishing, Paris.
DOI: http://dx.doi.org/10.1787/9789264119925-en.

OECD (2010), "Interim Report of the Green Growth Strategy: Implementing our commitment for a sustainable future: Meeting of the OECD Council at Ministerial Level", 27-28 May, OECD internal document.

OECD (2009), *Strengthening Regional Fisheries Management Organisations*, OECD Publishing, Paris.
DOI: http://dx.doi.org/10.1787/9789264073326-en.

OECD (2006a), *Using Market Mechanisms to Manage Fisheries: Smoothing the Path*, OECD Publishing, Paris.
DOI: http://dx.doi.org/10.1787/9789264036581-en.

OECD (2006b), *Fishing for Coherence: Fisheries and Development Policies*, The Development Dimension, OECD Publishing, Paris.
DOI: http://dx.doi.org/10.1787/9789264023956-en.

OECD (2003), *The Costs of Managing Fisheries*, OECD Publishing, Paris.
DOI: http://dx.doi.org/10.1787/9789264099777-en.

Seafish (2008), *Work currently being undertaken by Seafish to improve fuel efficiency in the fishing fleet*, Factsheet/Datasheet, April.

Sumaila, U.R. et al. (2012), "Benefits of rebuilding global marine fisheries outweigh costs," *PloS one 7.7*: e40542.

World Bank (2013), "Fish to 2030: Prospects for fisheries and aquaculture", *Agriculture and Environmental Services Discussion Paper*, No. 3, World Bank Group, Washington D.C.
http://www.dfo-mpo.gc.ca/aquaculture/sustainable-durable/index-eng.htm;
http://www.akvaplan.niva.no/en/resource_centre/news_detail/maximus_a_new_european_funded_r4smes_project.

World Bank (2009), *The Sunken Billions*, The World Bank Group, Washington, D.C.

Chapter 2

Green growth in fisheries

This chapter will follow the steps of the OECD Green Growth Strategy (GGS), taking each in turn and applying it in the context of fisheries. It shows how fisheries can contribute to the overall GGS for a country, and how the GGS as a strategy can be applied to fisheries itself. It emphasises the need for a strong, science-based approach to stock management as the foundation of resource sustainability, combined with a transparent and reactive policy development cycle to ensure that fisheries deliver maximum possible benefits.

The OECD Green Growth Strategy (GGS) can help fisheries sustainably contribute to economic growth, regional and rural development, and environmental objectives such as reducing climate change emissions. The GGS offers a structured approach to help policy makers identify and overcomes current problems and establish a policy development process that is focussed on achieving clear objectives through pragmatic reforms.

Build a green growth model

The first step in devising a strategy for green growth is developing an overall vision for the fisheries sector. Identifying current problems and risks is the first step in demonstrating the potential gains from improvements. Overfishing and illegal fishing, overcapacity, waste and inefficiency are problems faced by all countries to some extent, and this section will discuss the costs of these problems. As well, new sources of green growth can be identified that can help policy makers quantify the benefits of reform and persuade stakeholders of the value of change.

The risks of business as usual

Overfishing

Overfishing reduces the long-term productive capacity of the resource, lowering the potential benefits that fishers, consumers and societies at large can derive from it. In order to understand why overfishing continues despite the broad recognition of its negative impacts, the incentives and institutions of the fishery must be understood. For many, the fisheries problem is about co-ordination, as the incentives faced by fishers as individuals act at cross-purposes to the greater good. This has been called the common good problem.

But the problem is also one of institutions and the economic and social infrastructure that underpins the fishery. When fisheries managers are unable to set or enforce restrictions on fishing that would ensure sustainable exploitation of the resource, the problem lies in the ability of institutions to take all forms of information into account in decision making, the influence of pressure groups and election cycles and a lack of useful feedback when bad outcomes occur. When scientific information is uncertain or not trusted, it can be replaced by negotiation. Some governance structures are more adapted than others to take the long-term view that is essential for sustainable exploitation of the resource.

There is broad international agreement on the need to improve fish stocks. The WSSD Johannesburg 2002 Plan of Implementation called on governments, inter alia, to "maintain or restore stocks to levels that can produce the maximum sustainable yield with the aim of achieving these goals for depleted stocks on an urgent basis and where possible not later than 2015". A more recent call is the Aichi 2011 Strategic Plan for Biodiversity calling for fisheries to be sustainably managed by 2020 (Box 2.1). The work of the COFI and the FAO Code of Conduct for Responsible Fisheries provide important inventories of practical, evidence based "ways forward" to achieving green growth in fisheries.

Box 2.1. The Aichi Strategic Goal B Target 6

By 2020 all fish and invertebrate stocks and aquatic plants are managed and harvested sustainably, legally and applying ecosystem based approaches, so that overfishing is avoided, recovery plans and measures are in place for all depleted species, fisheries have no significant adverse impacts on threatened species and vulnerable ecosystems and the impacts of fisheries on stocks, species and ecosystems are within safe ecological limits.

Source: Convention on Biological diversity, http://www.cbd.int/sp/targets/.

Illegal, unreported and unregulated fishing

Illegal, unreported and unregulated (IUU) fishing is a practice that also contributes to considerable risk of depletion of certain species as well as damage to ecosystems (Box 2.2). The UN conference on sustainable development Rio +20 in June 2012 reiterated the commitment of the parties to eliminate IUU fishing activities, acknowledging that they undermine the sustainable use of fisheries. The Introduction to the *FAO International Plan of Action to Prevent, Deter and Eliminate Illegal, Unreported and Unregulated Fishing* notes that

> "the issue of illegal, unreported and unregulated (IUU) fishing in world fisheries is of serious and increasing concern. IUU fishing undermines efforts to conserve and manage fish stocks in all capture fisheries. When confronted with IUU fishing, national and regional fisheries management organizations can fail to achieve management goals. This situation leads to the loss of both short and long-term social and economic opportunities and to negative effects on food security and environmental protection. IUU fishing can lead to the collapse of a fishery or seriously impair efforts to rebuild stocks that have already been depleted. Existing international instruments addressing IUU fishing have not been effective due to a lack of political will, priority, capacity and resources to ratify or accede to and implement them." (FAO 2001).

IUU fishing activities not only damage the environment and threaten biodiversity; they also undermine labour standards, harm markets for legally harvested fish, encourage corruption and reduce prospects for food security and economic growth and stability especially in developing coastal nations.

Recently, there has been a move to consider IUU fishing in the larger context of "fisheries crime". This view recognises that IUU fishing is usually associated with and indeed requires other illegal activities such as tax and financial crimes, abuse of labour including slavery and human trafficking and smuggling. Interpol has been engaged in this area under its Project Scale initiative.

Overcapacity

Overcapacity, when the capacity of the fleet is higher than that required to harvest the stock at the targeted level, can be a threat to the conservation and sustainable exploitation of marine biological resources. Overcapacity reduces average profitability for fishers and can lead to too much effort in the fishery. Overcapacity ties up capital and creates pressure on management to increase allowable harvest through lobbying and other demands for support. Excess capacity also increases the likelihood of IUU fishing as more vessels are available for this activity.

OECD member countries have put considerable effort into reducing excess capacity including through decommissioning schemes of different types. These efforts have not always been successful. In some cases capacity trickles back into the sector over time, or fishing effort was not reduced to the extent anticipated. That said, there are things governments can to do help address capacity problems and evidence that there are significant benefits to getting capacity-management policy right (OECD 2009) (Box 2.3).

Inefficient use of resources

A substantial share of fish resources goes unused. A portion of the harvest is lost as discards and never landed. After landing during processing, low-value trimmings for which no market exists are disposed of. Fish is frequently sold in a fresh and perishable form, and so some percentage is disposed of unsold at the retail level or thrown away as spoiled by consumers.

Discards and bycatch are fish that are caught but which are not desired or permitted to retain (Box 2.4). While there is certainly an economic component to discarding, in many cases policies are in place that either demand or prohibit discarding, or that change the economic calculus in deciding whether to discard or not. For example, high-grading is the practice of keeping only the higher-valued fish caught and discarding the rest. This behaviour depends on how quotas are allocated to fishermen, and the monitoring and enforcement of restrictions on such activity.

Box 2.2. What is IUU fishing?

"IUU fishing" is a broad term that captures a wide variety of fishing activity, most of which is illicit. Illegal fishing is, by definition, wrongful. Any fishing activity that should be reported but is not (or is misreported) is also wrongful. Although unregulated fishing may or may not be wrongful, depending on the circumstances, it can frustrate the achievement of sustainable fisheries.

The common thread is that IUU fishing may generally be said to occur in violation of - or at least with disregard for – applicable fisheries rules, whether adopted at the national or international level.

Illegal fishing refers to fishing activities:

(1) conducted by national or foreign vessels in waters under the jurisdiction of a State, without the permission of that State, or in contravention of its laws and regulations;

(2) conducted by vessels flying the flag of States that are parties to a relevant regional fisheries management organization but operate in contravention of the conservation and management measures adopted by that organization and by which the States are bound, or relevant provisions of the applicable international law; or

(3) in violation of national laws or international obligations, including those undertaken by cooperating States to a relevant regional fisheries management organisation.

Unreported fishing refers to fishing activities:

(1) which have not been reported, or have been misreported, to the relevant national authority, in contravention of national laws and regulations; or

(2) undertaken in the area of competence of a relevant regional fisheries management organization which have not been reported or have been misreported, in contravention of the reporting procedures of that organization.

Unregulated fishing refers to fishing activities:

(1) in the area of application of a relevant regional fisheries management organization that are conducted by vessels without nationality, or by those flying the flag of a State not party to that organization, or by a fishing entity, in a manner that is not consistent with or contravenes the conservation and management measures of that organization; or

(2) in areas or for fish stocks in relation to which there are no applicable conservation or management measures and where such fishing activities are conducted in a manner inconsistent with State responsibilities for the conservation of living marine resources under international law.[11]

Source: FAO (2002), "Implementation of the International Plan of Action to Prevent, Deter and Eliminate Illegal, Unreported and Unregulated Fishing", *FAO Technical Guidelines for Responsible Fisheries*, No. 9, Food and Agriculture Organization Publications, Rome.

Box 2.3. Addressing overcapacity in the European Union

Since the 1990s the European Union has shifted away from expanding the EU fishing fleet and focussed public support on adapting it to available natural resources. Between 2007 and 2013, approximately EUR 840 million has been invested in reducing the EU fishing fleet by 4 000 fishing vessels. More generally, this has led to an overall reduction of the EU fleet by more than 25% between 2000 and 2014. The process of combining long term management plans, establishing fishing opportunities in line with the MSY objective and the decommissioning of fishing vessels have contributed to the reduction of pressure on fishing resources. This has contributed to the first hopeful signs of a fisheries recovery: fishing at sustainable levels is smarter economically as it helps maximize profit and income for the sector which in turns contributes to improving the welfare of coastal communities.

At a global level, more is needed to achieve the right balance between fishing capacity and resources. Although there have been significant developments in recent years to improve international rules and processes, too often their principles and general obligations are not translated into reality. Compliance still leaves too much to be desired and violations are hardly sanctioned. While some battles are being won, such as the increasing number of stocks exploited at MSY levels in the Atlantic, the North Sea and the Baltic, others still need to be fought such as IUU fishing worldwide, as well as large amounts of public resources being allocated on harmful capacity enhancing subsidies.

Rules and guidance are in place, but more political will is needed to enforce them. The FAO Action Plan on overcapacity and joint recommendations of the Regional Fisheries Management Organisations for tuna (Kobe) for capacity reduction and transfer are a step in the right direction but need concrete implementation.

It is important to have tools to assess and monitor worldwide capacity, like a global record of vessels and a single system of vessel identification. Enforcement measures and sanctions also have a role to play, as do voluntary joint initiatives, along the lines of what has been done to curb illegal fishing. These specific tools should be part of a coherent approach for each ocean fisheries that includes development support and trade measures.

Source: DG MARE, European Commission.

Box 2.4. Discards waste resources and complicate management

Discards, the proportion of total catch that is returned to the sea (in most case dead, dying or badly damaged), represent a significant part of the world's marine catches and is generally considered a wasteful misuse of marine resources. The first global assessment was published in 1994 and it identified a total discard of 27 million tonnes (Alverson et al., 1994). The latest global study conducted by FAO in 2005 suggests that discard have dropped to 7.3 million but the figures are not totally comparable. Even if the first was overestimated and the latter underestimated, reductions seem to have been significant. The latest assessment corresponds to a weighted global discard ratio of 8%. However, large variations among fishing methods and regions exist (Kelleher, 2005).

Source: FAO (2011), *Global Food Losses and Food Waste*, Food and Agriculture Organization Publications, Rome.

Discarding can be considered in many cases a sub-optimal use of resources. Beyond the direct physical waste of fish biomass, discarding involves inefficiencies in production as effort is expended in the catching, sorting and disposal of discards. Discards complicate the measurement of fishing effort and mortality, as it makes harvest different from landings. This can lead to a range of undesirable environmental effects on the target species, on endangered species and on the ecosystem. An estimated 8% of global marine catches are discarded, equivalent to 7 to 8 million tonnes per year (Kelleher 2005). Discard rates (the proportion of the catch discarded) vary widely by type of fishing gear. Shrimp and demersal finfish trawls account for 50% of global discards. Small-scale fisheries and passive gears generally have lower discard rates than industrial fisheries with active gears. Discards have reduced in recent decades due to more effective fisheries management, selective gear use and better fisher practices in OECD countries and the use of bycatch for aquaculture feeds in developing countries.

There are many risks in business as usual, and the status quo does not look sustainable in many ways (Figure 2.1). Fortunately, there are solutions to all of these risks and other factors that cause fisheries to be less productive, profitable and sustainable than they can be. These risks are also opportunities to bring new benefits and growth from the fisheries resource. The next section discusses the potential upside to reforms in fisheries that either address these risks or bring new sources of sustainable growth to the fisheries sector.

Figure 2.1. Risks of business as usual in fisheries

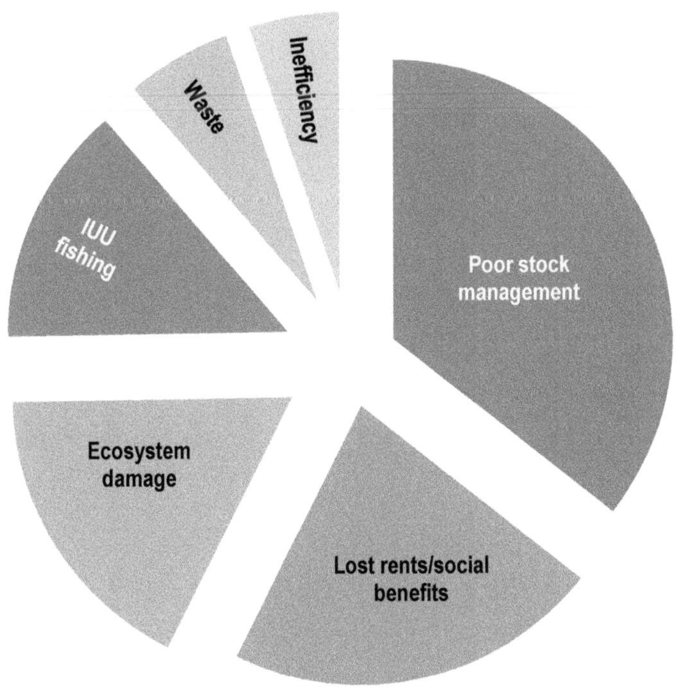

The potential in green growth

Better resource management

In many jurisdictions, depleted fisheries have been or are being rebuilt. Stakeholders are more aware than ever of the limits of the resource, consumers are demanding sustainability, and governments are acting to reform old approaches. For example, actions taken in the United States guided by the Magnuson-Stevens Act have led to 28 fisheries being rebuilt, supporting as a result more fishing activity worth an additional USD 585 million per year (NRDC, 2014). Global rebuilding could add as much as USD 50 billion in income worldwide (World Bank, 2009; Sumaila et al., 2012).

Improved management of stocks has been shown to increase profits through higher catch per unit effort (CPUE). A study by the New Economic Foundation (NEF) suggests that in the United Kingdom "*To land the same quantity of fish as they did in 1889, UK trawlers must now exert 17 times more effort – equivalent to a 94% fall in productivity*" (Crilly and Esteban, 2011). It has also been suggested that restoring 39 North East Atlantic stocks could deliver up to GBP 14.62 billion per year in gross revenues (Crilly and Esteban, 2012).

Fisheries management has evolved. The transition from open access fisheries to limited entry is essentially complete, at least in developed countries' exclusive economic zones (EEZs). Currently, fisheries management systems are more capable than ever to control harvest, ensure sustainability of the fish stock and share it among different users. At the international level, guidelines on best practices and agreements such as the *FAO Code of Conduct for Responsible Fisheries* (FAO, 1995) are slowly but inevitably bringing improvements (Figure 2.2). The changes required to establish "modern" management systems that take broader concerns into account are also increasingly in evidence. Modern fisheries management moves beyond simply ensuring sustainability (though this remains the essential core of management) to defining the role that fisheries should play in the economy and society and management approaches that will bring about more benefits at lower costs, both for fishers and for governments.

Figure 2.2. Historical progression of fisheries management

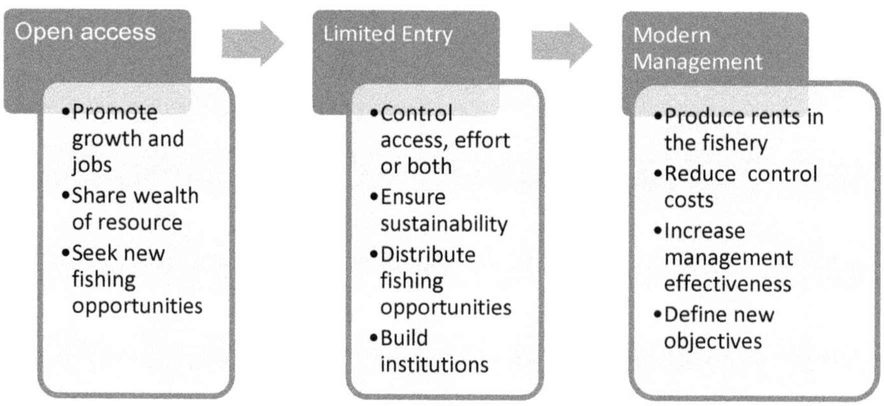

Considerable growth potential can come not only from producing more fish, but by earning more income from each fish harvested. Pursuing lower costs and higher value-added is another way to generate economic growth from fisheries while still respecting natural limits. One of the best ways to do this is through stock rebuilding. Higher populations of fish mean that it is easier to catch the allowed harvest, lowering costs. Improving the age-structure of the stock by maintaining healthy populations can also increase average sized fish, which are generally more valuable.

Generally speaking, most domestic fisheries taking place in the exclusive jurisdictions of countries are not at immediate risk of collapse, even if they are not producing at their maximum potential. High-seas fisheries – that is, those beyond EEZs (generally 200 nautical miles off a coast) – face more pressures as they are exploited by fleets from many countries as well as significant illegal, unreported or unregulated (IUU) fishing. In these fisheries, improved technologies that increase the effectiveness of monitoring and enforcement and lower its cost can help.

Fishing is an important part of the rural economy, though the number of fishers and vessels has been declining over time in most OECD countries as fishing opportunities decrease and consolidation takes place (Figure 2.3). The average age of fishers is increasing and many will leave the industry over the next decade. The desire to preserve the role of fishing as a contributor to local economies, especially when these economies are disadvantaged or offer limited alternatives, has led to many policies aimed at maintaining the size and distribution of fishing fleets and ancillary activities such as fish processing. This policy objective is at odds with market based approaches such as individual transferrable quotas (ITQs), which tend to promote consolidation and rationalisation of fishing capacity. Part of the challenge for green growth is balancing social issues with the economic cost in terms of lost efficiency. Ultimately, new approaches will need to be found that eliminate this trade-off; the best solution is one that is economically efficient and socially acceptable in equal measure.

Figure 2.3. Number of fishers and fishing vessels, selected OECD countries 1999-2012

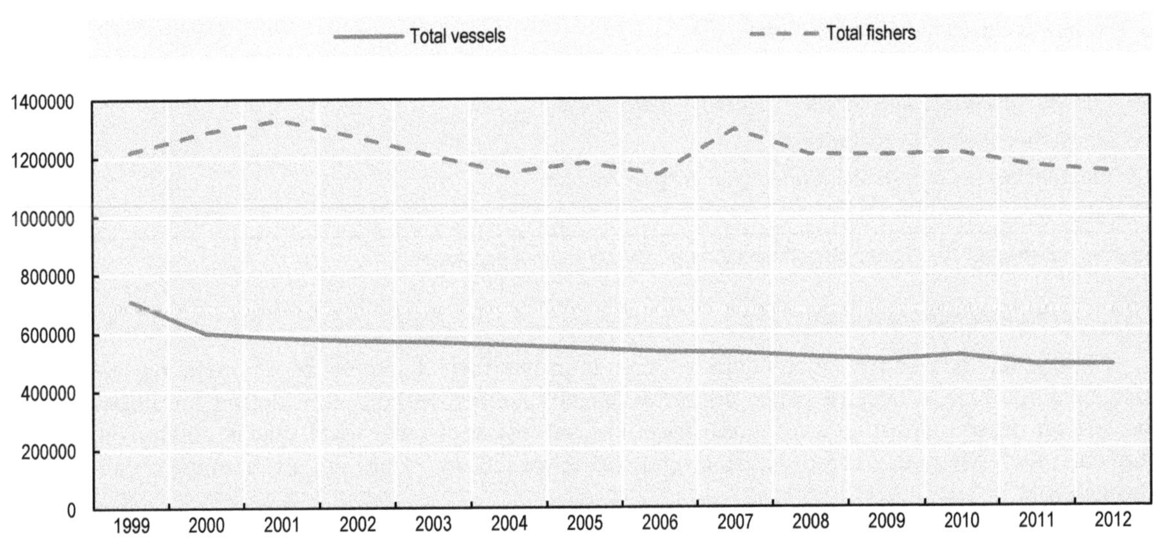

Note: Data not available for United States, Luxembourg, Hungary, Czech Republic, Slovak Republic.
OECD (2014), "Fisheries: Fishing fleet" and "Fisheries: Employment in fisheries", *OECD Agriculture Statistics* (database).
DOI: http://dx.doi.org/10.1787/data-00218-en and http://dx.doi.org/10.1787/data-00219-en.

Ensuring that all fishing activity is legal and recorded can bring significant benefits in the form of increased government revenue, higher fish prices and better resource use. These benefits are estimated to be between USD 10 and USD 23 billion annually (Agnew et al., 2009). For example, according to Agnew (2010), licensing the illegal fleet operating in West Africa would raise USD 4 million of government revenue. Excluding illegal vessels would further maximise societal gains by generating an additional USD 71 million. Along the same lines, Bonini (2011) estimates that East Atlantic and Mediterranean tuna fishers would increase their net benefits by USD 18 million over five years if illegal fishing was eliminated and current catch levels were strictly enforced as this would restore the sustainability of the Bluefin tuna stock and allow fishers to command a higher price for their catch.

Reducing waste

Central to the idea of green growth is improving the efficiency of natural resource use to enable economic growth even when those resources are fixed in quantity. Reducing waste can be an important part of that increase in efficiency. The global value of waste attributable to the fisheries sector is estimated to be in the order of USD 100 billion per year and USD 45 billion if economic waste attributable to overfishing is excluded. An estimate of the potential savings which may be made by waste reduction would require further study, but could be in the order of USD 30 billion per year (Table 2.1).

Improved management holds the key to continued economic growth in fisheries, as well as sustainable and healthy marine ecosystems. But there is more that can be done. Innovations that reduce current sources of waste and inefficiency will bring positive results. In addition, innovations in new products such as functional foods and non-food uses can open new avenues for adding value from marine harvests (e.g. fish oil tablets). Better scientific understanding of fisheries and the surrounding ecosystems will reduce risk and improve management. Enhanced enforcement will reduce the problem of IUU fishing, leading to better economic, social and ecological outcomes.

Table 2.1. Potential gains from reducing waste in fisheries
USD billions per year

Waste segment	USD billion
A. Harvest waste	59.9
A.1. Marine capture (including lost rents)	56.1
A.2. Inland capture (including lost rents)	3.8
A.3. Aquaculture (negative externalities offset by positive)	0.0
B. Post-harvest waste and losses (processing, distribution)	28.5
C. Consumer and household waste	10.0
D. Externalities not otherwise included (including emissions)	6.3
Total	104.6

Source: Kelleher (2013), "Green Growth and Waste", OECD internal document.

International cooperation

Experience has shown that countries can improve their fisheries by cooperating on fisheries management. This is because when countries act unilaterally without considering the impact of their harvest on other users of the resource, they will harvest above the level that is collectively optimal (Levhari and Merman, 1980; Kaffine and Costello, 2011). That is, all countries gain through cooperation.

The traditional way that countries cooperate to harvest shared resources is through treaty-like agreements that share responsibility for stock assessments, agree on a harvest-rule for setting an overall TAC, and allocate fishing opportunities among countries. Allocations within countries are usually handled outside the agreement, so fishers from different countries targeting the same stock may operate

Figure 2.4. International Oceans Governance Structure

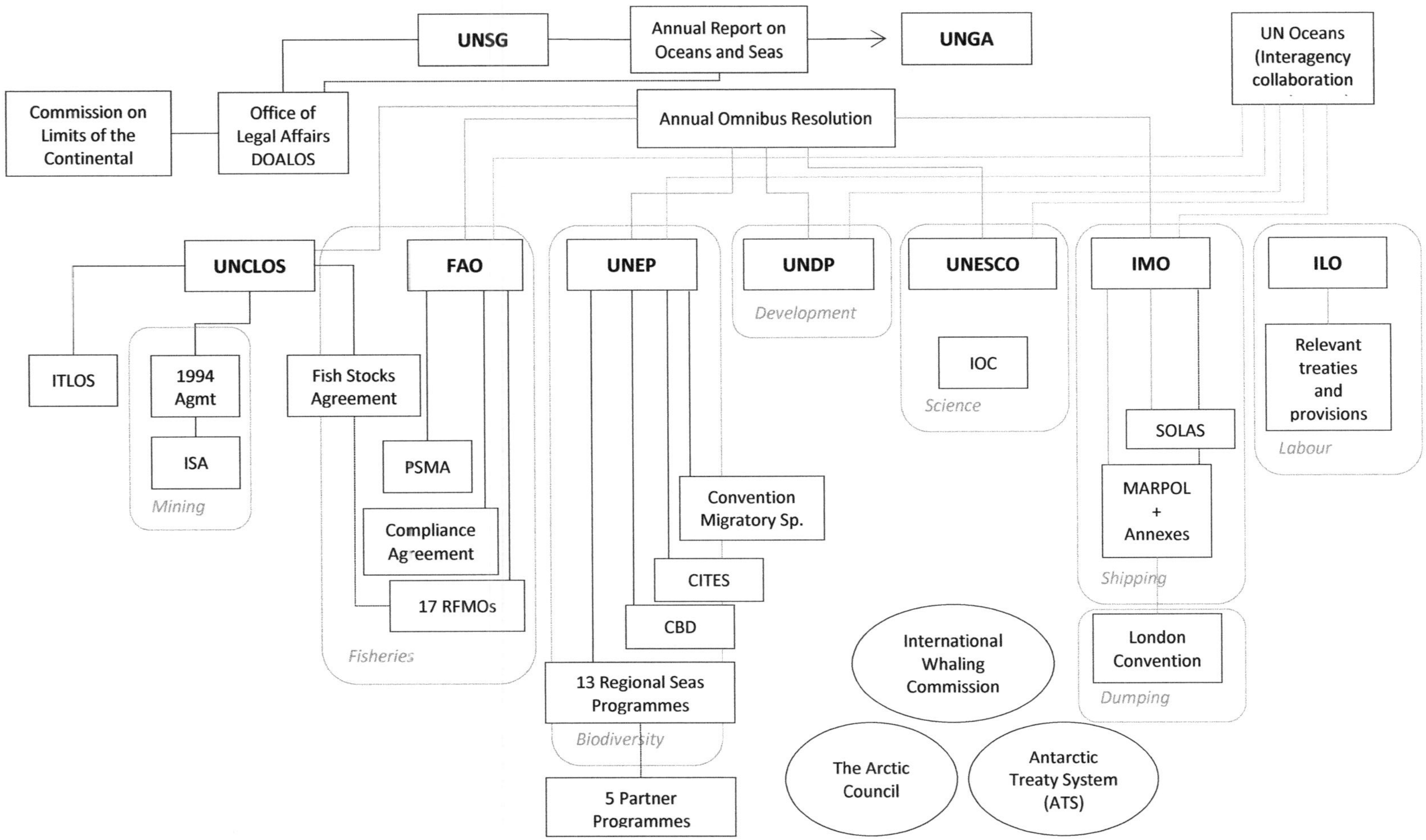

Source: Global Ocean Commission (2014) *From Decline to Recovery: A Rescue Package for the Global Ocean*, Global Ocean Commission Report 2014, Oxford.

under different regulatory systems and have different incentives. This approach depends on good stock assessments and harvest control rules and effective national MCS systems.

This approach is practical and has been effective in many contexts, but ignores the potential benefits of extending market approaches to the international context. For example, Norway and the EU cooperate in the North Sea via a substantial system of quota exchanges. EU vessels fish quotas for North-East-Arctic cod in the Barents Sea, in return for corresponding access for Norwegian vessels in EU waters. This arrangement originates in pre-Exclusive Economic Zone fishing patterns and the exchange is intended to maintain a balance in the fishing opportunities provided. As well, reduced domestic ownership requirements for fishing vessels can allow capital and expertise to flow more freely between countries.

International cooperation goes beyond mutually-agreed harvest limits and is perhaps most important in the area of MCS and prevention of IUU fishing. The FAO Port State Measures Agreement is a good example of how international cooperation in landing controls can help reduce IUU fishing. There are a number of international agreements related to fishing, the most important of which is probably the United Nations Convention on the Law of the Sea (UNCLOS), and the FAO Code of Conduct for Responsible Fisheries. The current system of ocean governance, which involves a large number of organisations with different powers and responsibilities, has been evolving over the last 20 years in response to a growing awareness that we need to exploit stocks sustainably, but there is still room for improvement (Figure 2.4).

RFMOs have the mandate to set management measures including catch and fishing effort limits and technical measures, as well as monitoring compliance with these management measures. Measures are binding and members rarely make use of objection procedures. Transparency and participation of observers have increased the accountability of RFMOs. Most RFMOs carry out performance reviews, incorporate new technologies such as electronic monitoring and are strengthening the role of scientific advice in decision-making. Continuing this process of improvement can bring big gains.

Improving global governance of the oceans can reduce many of the risks to fish stocks. New technologies that improve the ability to monitor and track fishing vessels, record and share information, and coordinate action will make international cooperation easier and more effective, both by reducing its cost and expanding the scope of what is possible. Investing in better science and better data and using the best available scientific advice for policy making can bring dividends by reducing risk to fish stocks and helping to maximize the potential of the resource.

Promote the transition

The previous section described the risks inherent in the status quo, and pointed to the potential benefits of green growth in fisheries. This section will address some of the actions that governments can take to begin the process of putting their fisheries on a greener growth path, focusing on removing some of the obstacles that can prevent positive change and facilitating the changes that are needed to put fisheries on a greener trajectory for long-term growth.

Remove barriers to green growth

Fisheries policy has expanded its focus from food, jobs and economic activity to rebuilding, sustainable use, reducing excess capacity and other pressures on the resource. Advances in fishing technology have transformed our perspective on fish stocks from being essentially unlimited to requiring careful management if overfishing is to be avoided. Yet, many policies have continued essentially unchanged over this period, despite the changing circumstances in the fisheries. Part of a green growth strategy in fisheries is looking at longstanding policies and approaches to see whether they address current and future needs, and reform them if that is not the case.

Over time it becomes increasingly more difficult to reform policies. They gain established constituencies of beneficiaries and institutional investments and expertise built up to support them can also be a barrier to change. Path dependence is a well-known phenomenon in policy design; new approaches tend to be built on established practices and are influenced by them. Policy stability is often cited as a desirable characteristic given that fishers make long term investments that are contingent on the policy environment.

This is why the OECD GGS puts an emphasis on a systematic policy review process that makes use of indicators of progress against clear objectives. The policy development cycle ideally constantly reassesses current approaches against current situations with a view to ensuring that policies are responsive to needs and are as cost-effective as possible. It sometimes seems the case in fisheries that significant reforms come only after a crisis; breaking that pattern with proactive change can avoid unnecessary risks and hardship.

There have been some important changes in fisheries such as the expansion of EEZs and the development of new management techniques that give hopeful signs for the future (Box 2.5). There is some evidence that as these techniques demonstrate their merits, they will be adopted in other jurisdictions.

Moving to a green growth path for fisheries is not just about identifying the best policies; it is also about finding a reform process that works. The process of adjustment is not easy. It creates winners and losers and demands everyone to take a leap of faith that untested policies will lead to a better future than established and trusted approaches. Making that leap requires confidence that the benefits are real, that the probability of achieving them is high, and that the individuals required to bear the cost of the transition are better off in the new regime.

OECD work has identified some key constraints to greener growth, dividing the problem into low economic returns and low capacity for actors to capture potential economic returns (Figure 2.5). There are many ways that governments can overcome these constraints; the OECD GGS focuses on the opportunities provided through the power of markets, globalisation and innovation. Moreover, the OECD GGS emphasises the benefits that come from being the first to tackle green growth. Being an early adopter of Green-growth policies can lead to a first-mover advantage in green markets, a comparative advantage in new technologies, and an earlier return on investment in positive change.

Figure 2.5. Green growth diagnostic

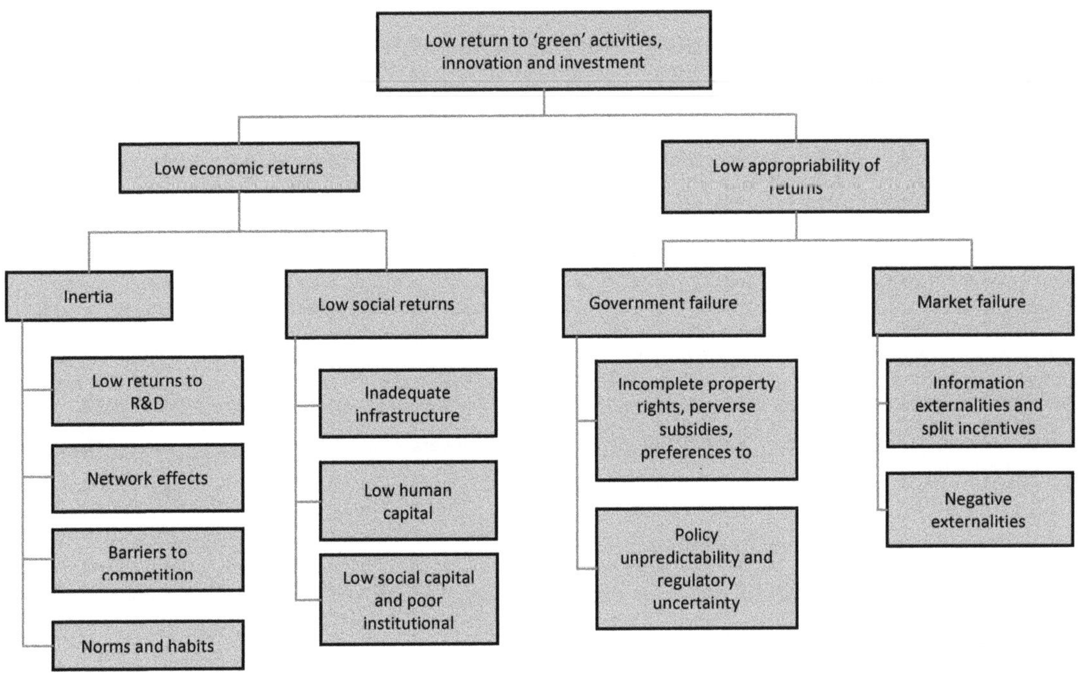

Source: OECD (2011), "Towards Green Growth: A Summary for Policy Makers", brochure prepared for the OECD Meeting of the Council at Ministerial Level, 25-26 May 2011, Paris.

Box 2.5. A brief history of fisheries management

Management of commercial fisheries, particularly in developed nations, has undergone a transformation since the extension of the exclusive economic zones (EEZs) of coastal nations in the late 1970s. This led to the enclosure of most of the world's productive coastal shelf habitat as the territory of sovereign nation-states, providing the opportunity to abandon economically inefficient and biologically unsustainable open access institutions (Gordon, 1954; Hardin, 1968). Nevertheless, the subsequent history of fisheries management has been a chequered one.

- Many nations, motivated by goals of biological sustainability and maximum sustainable yield, established seasonal total allowable catches (TACs) for target species and then sought to enforce these limits by a combination of controls on outputs (via monitoring of cumulative catch within the season) and inputs (e.g. limits on season length, vessel length, horsepower, etc.).
- These "regulated open access" institutions did provide some degree of protection against biological overexploitation. However, they did nothing to check open access competition for the TAC, fuelling a "derby" in which fishermen fished far too intensively and made wasteful investments in vessel capital to out-fish their competitors.
- The result was "cost side" dissipation of fishery rents through excess entry, overcapitalisation and overuse of variable inputs within-season, "revenue side" dissipation of rents due to derby-induced gluts forcing product into low-value markets (Homans and Wilen, 2005; Homans and Wilen, 1997).
- Modifications to combat excessive effort through buybacks and limited access programs to reduce active vessels have often yielded little economic benefit (Wilen 2006; Wilen 1988) as remaining vessels continue to face incentives to dissipate economic value in the pursuit of a larger share of the catch.
- Furthermore, the theoretical ability of such systems to limit biological exploitation have been hampered by poor enforcement of TACs under intense derby fishing, pressure to set high TACs as a short-run palliative to economic dissipation, and poor stock assessments – yielding a legacy of both biological and economic failure (Beddington, Agnew and Clark, 2007; Hilborn et al. 2003; Worm et al. 2009).

A gradual paradigm shift has occurred in many jurisdictions in response to these shortcomings, viewing many of the negative symptoms of conventional management as arising from a common cause: the lack of secure property rights to the fish stock and its harvest (Arnason, 2012b; Grafton et al., 2006). Solutions to this problem, passing under the rubrics of "rights-based management", "market-based management", "incentive-based management" or "catch shares", address this lack by allocating shares of harvest or habitat to individuals or groups.

- Individual transferable quotas (ITQs) guarantee quota holders the durable use right to a share of the seasonal TAC and allow quota holders some flexibility to lease or sell this right.
- Harvester cooperatives (Deacon, 2012) allocate a share of the TAC to a group of fishermen and then allow fishermen within the cooperative to allocate and manage the quota in their collective interest.
- Territorial use rights fisheries (TURFs) (Cancino, Uchida and Wilen, 2007; Wilen, Cancino and Uchida, 2012) guarantee groups of users rights to utilize resources contained within well-defined coastal spatial areas.

While often stopping far short of the exclusivity, security, durability, and transferability associated with "strong" property rights (Arnason, 2012b), these approaches can eliminate much of the wasteful common-pool competition, leading to a slower pace of fishing, reversals of overcapitalization and input stuffing, and incentives to maximise the value of catch rather than merely its volume. The economic track record of rights-based tools, with widespread adoption in a significant proportion of OECD countries, has been very strong (Abbott, Garber-Yonts and Wilen, 2010; Grafton et al., 2006; Hilborn, 2007b; Hilborn, Orensanz and Parma, 2005; Leal, 2005) – dramatically enhancing income generation within the biological constraints of annual TACs. Furthermore, there is evidence that rights-based approaches may also further ecological sustainability goals (Branch 2009; Chu 2009; Costello, Gaines and Lynham, 2008; Essington, 2010; Essington et al., 2012; Melnychuk et al., 2012) – perhaps attributable to better attainment of management targets due to the measured rate of catch and enhanced stewardship incentives on the part of fishermen as "shareholders" in the future health of the fish stock. While not a panacea for all that ails fishery management (Degnbol et al., 2006; Ostrom, Janssen and Anderies, 2007), rights-based approaches are a valuable tool for overcoming the incentive failures that undermine status-quo approaches to management (Beddington, Agnew and Clark, 2007; Worm et al., 2009).

Source: Abbot (2013), *Integrating Recreational Fisheries into Fisheries Management: Challenges and Opportunities*, OECD internal report.

Facilitate transformational change

What can governments do to put their economies on a green growth path? In fisheries, the role of governments goes beyond the setting of the regulatory and policy environment to include direct management of the fish stock and its harvest. Public institutions must manage fisheries and ocean resources in trust for society. The fish stock is a public asset, and like any asset should be managed to maximize return while managing risk.

Since the government determines so much of the outcomes in fisheries, improving governance is one of the most direct and effective ways that governments can encourage green growth. The following sub-section describes some principles for institutional design that can help ensure that institutions are organised and function as well as is possible.

Governments also have a role to play in helping economic agents innovate and improve their operations, by removing perverse incentives, fixing externalities and promoting public goods. In the subsections that follow, the potential for improvements in the handling and use of fish materials, energy use, and the use of market approaches to management are discussed.[1]

Improve governance

The way governments organise themselves to make decisions and manage fisheries can have an important influence on how efficiently resources are used, how effectively objectives are reached and the health of the marine economy. Governance of fisheries is growing more complicated as the marine economy has evolved and the number of users and concerned parties has expanded. The OECD GGS asserts that co-ordination across different arms of the government is necessary to ensure progress in one area does not come at a cost to another. Such policy coherence is a necessary requirement to getting the most from the natural resource base.

Fisheries management is broadening its scope to be more policy coherent. This process is difficult and complex, but the rewards can be large. Many fisheries are grappling with moving from a single-species approach to managing multiple species—taking into account predator-prey relationships, for example. Ecosystem-based management is also gaining ground; this management approach tries to understand and account for the impacts of fishing on the broader environment and not just commercial species. Ultimately the larger marine economy as well as the ecosystem—the "blue economy" will be treated in a comprehensive and co-ordinated manner.

For example, the *Reykjavik Declaration on Responsible Fisheries in the Marine Ecosystem*, signed in 2001, declares that responsible and sustainable fisheries in the marine ecosystem require incorporating ecosystem considerations into management. Noting the challenges inherent in doing this, signatories recognised that work must be done to advance the scientific basis for incorporating ecosystem considerations, building on existing and future scientific knowledge (FAO, 2001b). Another example is the EU Blue Growth Strategy, a long term effort to support sustainable growth in the marine and maritime sectors as a whole. It is the maritime contribution to achieving the goals of the Europe 2020 strategy for smart, sustainable and inclusive growth (European Commission, 2012).

While it is recognised that policy-coherence requires a broad perspective and an inclusive approach across currently separate policy domains, it is also clear that sustainable management requires a strong focus on the health of fish stocks and simple rules to determine harvest. To reconcile these two requires a well-designed institutional framework. However, there is no such a thing as an optimal institutional design. The context in which institutions operate is too variable, and there are too many implicit trade-offs in institutional design for there to be a single best approach. All that can be done is to compare the characteristics and performance of different approaches.

There are a number of governance models in place in the fisheries sector, from highly centralised to highly decentralised, with related management tools. Decision-making can be concentrated in a single body such as the national legislature, or can be delegated to agencies, devolved to local levels, or made implicit in laws or regulation. Associated with each decision-making level are methods of resource allocation, from overall quota-setting to allocation by agencies, communities or markets. Together these describe the degree of decentralisation of management and institutions.

Because of the need to consult and cooperate with stakeholders, it is now generally recognised that highly centralised institutions do not work well in fisheries. So the question is what is the right degree of decentralisation, and decentralisation of what? Every level and type of institution, from legislature to bureaucracy to community to market has its particular strengths and weaknesses, so decentralisation is not simply a matter of picking the proper level to house all the machinery of fisheries management.

An efficient decision-making process involves keeping some responsibilities centralised while devolving others to subsidiary bodies better suited for them. For example, strategic decisions on fisheries often involve a number of levels and stakeholders. This can lead to balanced decisions as a number of opinions will be taken into account, but can also be a lengthy process. However, once decisions are taken, measures such as Harvest Control Rules (HCR) can be adopted where the level of exploitation of the stock is set every year by the rule via a scientific and administrative process. Lengthy discussions are avoided and management objectives are met (Figure 2.6). Another example is the EU adoption of multiannual plans. These provide a longer term perspective for stock management which then form the basis for deciding fishing opportunities on a yearly basis.

Stock management in fisheries is more likely to be successful when it focusses on harvest-control rules based on best available science. Fisheries managers who are asked to manage fish stocks while simultaneously managing fleet composition and distribution, rural employment opportunities, supply for processors and other matters will struggle to succeed at any of these tasks. The Tinbergen Rule holds that each objective should have its own tailored policy solution (Knudson 2008)[2]

Allocation and distributional issues primarily concern stakeholders, and allowing sector participants to directly control these decisions can promote co-operation and ease management. Regional or community bodies might be best placed to handle questions of allocation of fishing rights, or oversee market-based trading systems (Box 2.6). Regional bodies have grown in prominence in recent years, and new methods of community-based fishing continue to be proposed, such as Angling Management Organisations (Sutinen and Johnston 2003). Advisory Councils in the European Union are taking on an increasingly important role in the fisheries management system of the European Union (Box 2.7).

Figure 2.6. An example of a delegated system

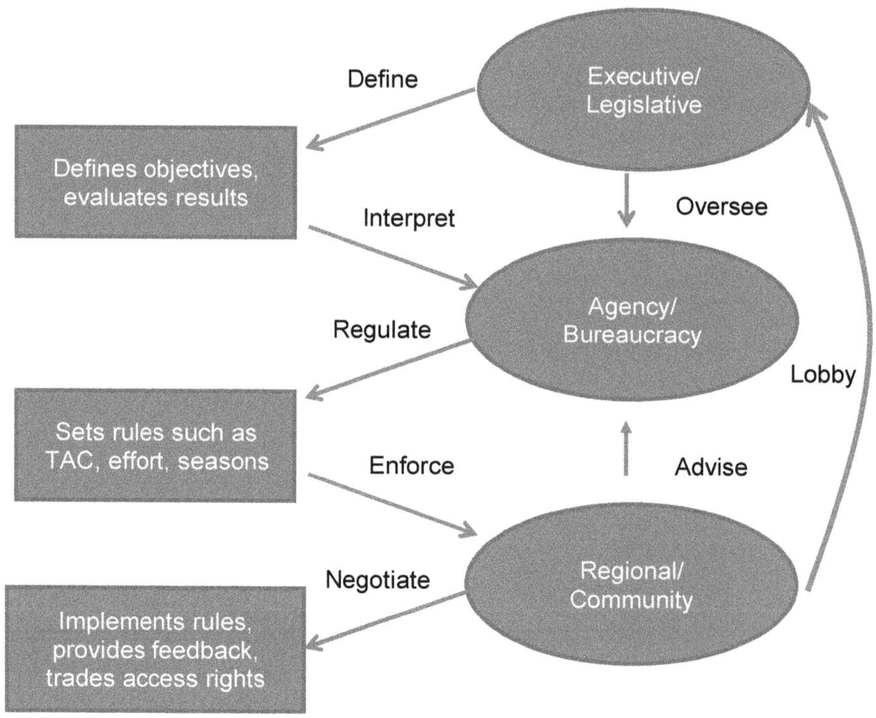

Box 2.6. Community-Based Fisheries Management in Korea

Community-based fisheries management (CBFM) was introduced in Korea in 2001. Fishers are the partners and initiators of management actions for their fisheries, in addition to the already-established rules and regulations for the sustainability of their local fisheries.

Under the CBFM, fishers' groups take voluntary management measures for their own fisheries, and actively participate in the decision-making process for dispute settlement; income generation, fishing ground and resource management, and stock enhancement in the framework of relevant fisheries laws and regulations. An increasing number of fishers' groups are joining the CBFM.

Source: OECD (2013), *OECD Review of Fisheries: Policies and Summary Statistics 2013*, OECD Publishing, Paris. DOI: http://dx.doi.org/10.1787/rev_fish-2013-en.

Box 2.7. Advisory Councils in the European Union

The 2002 reform of the CFP introduced in the decision making process Regional Advisory Councils (RACs), that is stakeholders' organizations bringing together the industry and other interest groups affected by the Common Fisheries Policy, such as environmental NGOs and consumers' organizations, with a view to advising the Commission on matters of fisheries management in respect of certain geographical areas or fields of competence. There were indeed five RACs of regional character (Baltic Sea RAC, North Sea RAC, North Western Waters RAC, South Western Waters RAC and Mediterranean RAC) and two RACs of a more general scope (the Pelagic RAC and the Long Distance RAC). RACs contribution to the achievement of the objectives of the CFP has been very constructive.

With the new CFP reform, these stakeholders' organizations are transformed to Advisory Councils (ACs) and their role is reinforced. The Advisory Councils may, in particular, submit recommendations on matters relating to the management of fisheries and the socio-economic and conservation aspects of fisheries and aquaculture to the Commission and to the Member States concerned, inform the Commission and the Member States concerned of problems relating to the management and socio-economic aspects of fisheries and aquaculture, propose solutions to overcome those problems and contribute, in close cooperation with scientists to the collection, supply and analysis of data necessary for the development of conservation measures. In addition, within the framework of the new regionalized CFP, Member States shall consult the Advisory Councils when formulating joint recommendations for achieving the objectives of the Union conservation measures that they recommend. In addition to the above mentioned seven existing Advisory Councils, the new CFP foresees the creation of four new Advisory Councils for the Black Sea, Aquaculture, Markets and Outermost regions.

Source: Nicholas Dross (European Commission), personal communication.

A challenge is to maintain the independence of different institutions to prevent unhelpful political and lobbying pressure, while allowing feedback and cooperation to ensure that different levels work together effectively. The value of such communication was demonstrated in the 1992 collapse of the Atlantic cod stocks in Canada. Scientific assessments of the stock were carried out by government scientists who did not involve or inform fishers. As a result their estimates were mistrusted, intensifying sector lobbying pressure on the government to maintain or increase quotas (Shelton, 2007). Once scientists allowed fishers to participate in sentinel surveys and other data collection processes, the level of confidence in (and accuracy of) the results increased.

Good institutions tend to reflect three characteristics: **good availability of information** about the actions engaged in or supervised by the entity, **transparency in the decision-making process**, and **accountability in decisions** made and on enforcement measures taken. These are of particular significance in institutions like the ones involved in the management of fisheries because of their intermediate status between policy makers and actors. They are also essential to successful institutional reform as meeting these standards dispel suspicion and promote trust in and cooperation with institutions.

Gaining stakeholder cooperation requires investing in developing beliefs, values and norms of behaviour that can support and legitimize new rules and institutions. Without this, decisions tend to be challenged, disrespected, openly disobeyed or interfered with politically.[3] Fishers need to believe that the behaviour demanded of them by the management system is also the right thing to do. Consultation and involvement in the process are essential parts of this, but to change norms of behaviour can also require time for the benefits of new systems to become clear (and for new constituencies of beneficiaries to appear).

A significant challenge for fisheries management has been the practical difficulties in measuring the size of the fish stock and predicting its response to fishing effort. Uncertainty around the science and data can lead to information being replaced by negotiation in the decision-making process. To be effective, institutions should be able to incorporate information from different sources, evaluate the relative quality of that information, set goals that are based on evidence, and be adaptable when facts change.

Transparency has to do with the procedures for making decisions and for implementing the rules defined either at the broad institutional level or at the level of programme implementation. There is a difficult trade-off between strictly detailed procedures which are easy to communicate and control but introduce a lot of rigidities that may significantly increase costs (including long delays, interminable negotiations, repeated votes) and flexible rules that facilitate adaptation but could allow the decision process to become arbitrary. Public access to procedures, including for parties not directly involved, is one way to reduce the sharpness of this trade-off by making rigid systems run more smoothly and flexible ones less arbitrary. Consider on the other hand the undesirability of decisions made through late night negotiations or behind-the-scene lobbying of special interests.

Accountability on decisions made and on enforcement measures taken is also crucial to the legitimacy of institutions. This has *ex ante* and *ex post* elements. At a time a decision is made, it must be justified, so that the responsibilities in decision-making are clearly identified. The different contributions of the involved parties should be identifiable. For example, experts who delivered reports that supported decisions, criteria used by public servants to evaluate alternative solutions, and so on.

Accountability *ex post* requires that an assessment be made of results. This is an element the process of continuous policy review and improvement called for in the OECD GGS. The GGS emphasizes the value of indicators and evidence for successful institutional design and reform. Defining success though indicators moves the basis of evaluation of management actions from process to results. It should never be the case that all parts of the management system are considered to have done their job properly, yet the fishery suffers a bad outcome. Except in rare cases of external causes, it should be possible to define success or failure at the institutional level and identify remedies when the latter occurs.

Table 2.2. Elements of good governance

Essential characteristics
Good availability of Information
Transparency in the decision-making process
Accountability in decisions
Desirable Characteristics
Capacity to coordinate and delegate across institutions
Clear assignment of rights and responsibilities
Involvement of stakeholders
Methods of appeal
Capacity to promote good norms of behaviour

Source: Menard, C. (2014), "Institutional Aspects of Governance in Fisheries Management", OECD internal document.

There must be mechanisms of appeal available through which those who disagree with a decision can challenge it. These mechanisms must meet two conditions. First, the set of decisions that can be appealed and under what circumstances needs to be delineated in order to avoid repeated challenges by small minorities or by eternal challengers, which could make the process intractable. Second, the entity in charge of examining appeals must be independent of the micro-institution that is responsible for making or implementing decisions. Three possible mechanisms of appeal are commonly used: mediation, based on the common agreement to look for an acceptable settlement (e.g. through a conciliation commission); arbitration, which transfers decisions to independent parties; and courts. Costs tend to be higher for courts than for arbitration, and higher for arbitration than mediation.

To summarise, institutions should be designed to use and share the best possible information and be transparent and accountable. Other characteristics of institutions that are generally desirable are the capacity to coordinate and delegate, clear assignment of rights and responsibilities, involvement of stakeholders, methods of appeal, and capacity to promote good norms of behaviour (Table 2.2).

Example: Institutional design in the United States

The US Magnuson-Stevens Act (MSA) provides a good example of decentralisation using an agency approach. The Act is a piece of federal legislation that lays out the goals of fisheries policies. It identifies rebuilding and managing for maximum sustainable yield as key priorities and delegates' responsibility for meeting these objectives to the National Oceanographic and Atmospheric Administration (NOAA) through its sub-agency the National Marine Fisheries Service (NMFS). Substantial authority in turn is delegated to Regional Fishery Management Councils (RFMCs) who develop fishery management plans to restore and manage stocks. NMFS evaluates, approves, and implements these management plans. It therefore reflects in broad terms the structure shown in Figure 2.10.

An important part of the MSA is the formal role of data in decision making. Under the Act, stock assessments are used to trigger regulatory action (Box 2.8). This Act also requires rebuilding to take place within ten years, with some exceptions. While this has been criticised as inflexible, such a deadline has a number of important benefits. It allows compliance with the requirements of the Act to be clearly assessed, it makes the rebuilding target real and enforceable, and lowers the political transactions costs surrounding identifying the relative importance of stock rebuilding versus near-term economic losses.

Box 2.8. Rebuilding under the US Magnuson Stevens Act

A fish stock is considered overfished when it falls to a biomass or population level that jeopardizes the stock's capacity to produce maximum sustainable yield (MSY) on a continuing basis. In practice, the National Marine Fisheries Service and regional fishery management councils typically set an overfished "threshold" as part of designating what are called status determination criteria for the stock. When stock abundance falls below the threshold, the agency designates the stock as overfished and requires the relevant regional council to develop and implement a rebuilding plan as part of a fishery management plan or plan amendment. The rebuilding plan sets a rebuilding time period that meets the requirements of the law: "as short as possible,...not [to] exceed 10 years except in cases where the biology of the stock of fish, other environmental conditions, or management measures under an international agreement in with the United States participates dictate otherwise." The plan must also include management measures to constrain the fishing mortality level to rebuild the stock within this time period. When a population increases to above the overfished threshold but remains below the rebuilding target, it is considered not overfished but still rebuilding. By law, NMFS must review rebuilding progress no less than every two years; when the agency determines that inadequate progress has been made, the council is to be informed, and it must implement responsive management measures.

Source: NRDC (2013), *Bringing Back the Fish, An Evaluation of U.S. Fisheries Rebuilding Under the Magnuson-Stevens Fishery Conservation and Management Act*, Report # R-13-01-A, February.

Improving energy efficiency in the capture fisheries sector

Improving energy efficiency in fisheries serves two main purposes in helping the sector onto a greener path for growth. First, it reduces CO2 emissions, lowering the carbon footprint of fishing. This helps contribute to national commitments and objectives with respect to climate change. Second, it implies an increase in resource use efficiency that can also mean higher profits for fishers for a given harvest.

There are a number of ways that governments can support the change to more energy-efficient fishing. This has been the subject of some attention in past OECD work, including *Fuel Tax Concessions in the Fisheries Sector* (Martini 2012), *Green Growth and Energy Use in Fisheries and Aquaculture* (forthcoming) and *Energy Use in Fisheries: Policy Responses* (forthcoming). That last document investigates the role of government in promoting energy efficiency and informs the content of this section.

A consistent message in OECD work on energy use in fisheries is that energy use should be looked at as part of a larger strategy for the sector. That is, policies designed to improve energy efficiency should be coherent and compatible with other objectives for the sector. Another observation is that improved stock management that increases catch per unit effort (CPUE) tends to deliver the greatest improvements in energy efficiency, in addition to other benefits. As stock management is considered as part of the larger GGS for fisheries, this aspect of promoting energy efficiency is not further discussed in this section.

Many countries have national initiatives for increasing energy efficiency, and a first step should be ensuring that fisheries are fully integrated with these. National programs vary, but usually include, *inter alia,* purchasing guides or standards for energy-efficient systems, research and promotion of best practices, and support to efficiency-improving technology. Objectives for energy-efficiency improvements in fisheries should be set in the context of broader national objectives.

Subsidies and incentives are common policy tools used as part of national energy efficiency schemes. While there may be a role for such tools in fisheries, they must be approached with a heightened level of caution. That is because there is a risk that policies can aggravate existing capacity problems and reduce the capability of the fisheries management system to effectively control effort and harvest. A better approach may be to ensure that fishers are aware of and can access existing programs that are supportive of energy efficiency, such as R&D or investment credits and incentives in the tax code. Another option may be to make access to support policies contingent on fishers meeting a high standard of energy efficiency or participating in training or other programs designed to help fishers improve the efficiency of their operations.

Helping fishers to understand and plan improvements in their energy use through training, auditing and other informational services should be undertaken early on in the process. Governments can negotiate with the sector to establish voluntary standards and guidelines. Energy efficiency standards for different types of equipment can be established and the best performing technologies identified. Fuel meters and other technologies that offer feedback to fishers regarding their energy performance have a proven track record and should be promoted.

Since many of the actions that improve energy efficiency also improve profits, significant progress can be made without incurring additional net costs. But new approaches will not always be automatically adopted. Fishers may avoid changes in their operations when they are uncertain of the results, or when these changes lie outside their traditional skill sets. The incentives faced by fishers through short seasons or day-at-sea limits may lead them to use energy intensive approaches such as fishing faster or further from port, and so management reforms in some cases may be a prerequisite to significantly improve efficiency.

Setting the fishery on a path to green growth means getting the incentives right, and a big part of this is the price of energy. A fishery cannot be sustainably built on the premise that fishers can purchase fuel for less than the economic or social cost of using it. Therefore, making sure that the price of fuel is equal to its true cost and value to society is in important step. Governments should work to ensure that the price of fuel reflects its full external costs by using the principle of Pigouvian taxation: setting an energy tax equal to the external costs of its use. This is an application of the polluter-pays principle.

This also follows the commitment by G20 Ministers to eliminate fossil-fuel subsidies as stated in the 2009 Pittsburgh and 2010 Seoul Declarations.

As a practical matter, however, Pigouvian taxes are not always feasible to apply. The proper rate of such a tax can be difficult and contentious to determine, and if demand is very inelastic the required tax rate can be very high. Many governments have chosen instead to use a mix of instruments to promote fuel efficiency. Combining defined efficiency standards with mandates for progress and a reasonable tax on fuel can be a pragmatic and effective approach to improved energy efficiency.

Fuel tax concessions (FTCs) for fishers (and other sectors) reduce the cost of fuel relative to other sectors. These are put in place to ensure that certain energy-intensive fisheries segments continue to operate, to buffer the production impacts of increasing fuel prices or to reflect the fact that fuel taxes are a user fee for roads and other infrastructure. OECD Green Growth work has emphasised the need to eliminate environmentally-harmful subsidies, but it has also noted how difficult this is to accomplish. To make progress, a gradual approach is recommended, as is including other measures to "*solve delicate acceptability issues linked to redistributive or competitive impacts. This option requires judicious accompanying tools to address resistance due to a perceived increase in the tax burden and, of course, to handle the opposition of economic lobbies who will otherwise benefit from the exploitation of natural resources. Without accompanying measures, putting a price on carbon is creating losers, whose resistance is likely to block the implementation process or lead to exemptions that sharply deteriorate the overall efficiency of the measure*" (OECD 2013c).

Well-designed reforms can bring net benefits to the fishery by replacing existing policies with more cost-effective ones while avoiding negative side-effects. Evaluating the benefits and costs of FTCs and alternative approaches can help justify the need for reform. As well, a national approach to energy policy can be more feasible than single-sector reforms. This is because tax and energy policy go beyond the scope of fisheries, a consistent approach across sectors is more likely to be effective, and singling out one sector is less likely to be seen as fair when national energy policy is not itself well-designed.

Reducing waste in the capture fisheries sector

OECD Members have a wealth of experience in implementing a range of successful policies and programmes on bycatch and discards. Australia implements bycatch action plans for all Commonwealth (federal) fisheries and all OECD countries use a range of technical measures with regulations on gear, seasons, or area closures. New Zealand uses deemed values as economic incentives. Norway enforces a "no discards" policy and a similar policy is being phased in in the European Union. In Alaska, fishers help each other to identify areas with high bycatch in order to avoid early fisheries closures triggered by reaching a discard quota.

Many successful schemes depend on robust and representative information on discards or areas with high levels of undesirable bycatch. Alaska has 100% observer coverage in some fisheries, while Norway uses high levels of sampling to open and close fishing areas. Robust information on discard mortalities results in more accurate scientific advice. Placement of observers on smaller vessels is unrealistic and Denmark has fostered video monitoring of fishing vessels' catches combined with an accounting for discards as an element of individual catch quotas, i.e. the "quota" becomes a "mortality quota" rather than a "landings quota" (Dalskov 2010). The approach changes the incentives by placing a value on the discards and many of the technical issues have been resolved including automated video recognition and counting of discarded species.

Historically, fisheries in the European Union relied heavily on technical measures, often with inadequate knowledge of discards and poor observer coverage. Recently, the new CFP does away with the wasteful practice of discarding through the introduction of a landing obligation. This serves as a driver for more selectivity, and provides more reliable catch data. To allow fishermen to adapt to the change, the landing obligation will be introduced gradually between 2015 and 2019 for all commercial fisheries in European waters.

The FAO has described a range of voluntary and regulatory measures to address lost fishing gear (Macfayden et al., 2009). Market measures could also be envisaged, in particular application of extended producer responsibility (Ten Brink et al., 2009; OECD 2001). This is an approach which integrates environmental costs associated with goods throughout their life cycles into the market price of the products. It is commonly applied to recycling of batteries, vehicle tyres and electronic goods and can be mandatory, negotiated, or voluntary. Application to netting for fishing could potentially finance removal of lost gear and create incentives to reduce losses and develop more biodegradable gear. Linkages to ecolabelled nets made from recycled materials could also be considered.

Reducing waste in the fish value chain

Estimates of the amount of waste in the food chain are typically old and unreliable (for an exception see WRAP in the United Kingdom), so the first challenge is to quantify the problem in a manner consistent with OECD Council Recommendations (OECD, 2008). Physical accounts are needed, because knowledge of aggregate resource use and waste outputs is limited. Traditional monetary accounts and environmental statistics are also insufficient, as they lose track of the resource and waste flows, tend to track only the flows that are subject to regulation or classified as wastes requiring treatment, often aggregate waste products irrespective of origin, and may entirely miss major flows of materials that do not enter the economy at all, such as discards. Ideally, the physical, monetary and environmental accounts need to be linked and prepared for typical value chains to generate an understanding of the drivers and policy responses to waste. Countries would benefit from sharing experiences on fisheries waste measurement, on developing an understanding of drivers and impacts of policy responses, and on use of actionable indicators to track trends in waste.

The EU waste directive sets out general principles of waste reduction common to many public policy and green growth approaches by providing a generic, preferential hierarchy of choices: (i) waste prevention; (ii) re-use; (iii) recycling; (iv) other recovery (e.g. energy recovery); and (iv) disposal (i.e. dumping). Similar principles are echoed throughout the FAO Code of Conduct in relation to waste at different points of the value chain, in UNGA statements on oceans, in UNEP reports and numerous national environment policies and strategies.

All OECD members have waste policies or strategies and many have specific food waste reduction initiatives. However, few have comprehensive fisheries waste initiatives which cover the sector and the entire value chain. Nevertheless, most OECD members address priority fisheries waste streams as separate challenges (e.g. discards, lost fishing gear, processing waste). While there may be merit in bringing these discrete initiatives under a single notional waste umbrella, the diversity of actors, regulatory measures and technical solutions is likely to demand separate implementation of waste initiatives, leading fisheries authorities to prepare targeted guidelines for different waste problems. For example: SeaFish in the United Kingdom provides numerous guidelines; the US EPA even provides guidelines for recreational fisheries and Denmark for processing wastewater (Skall and Olesen, 2011).

A national fisheries waste initiative should build on existing actions (e.g. discard reduction programmes, ecolabelling schemes) and would ideally have several of the following elements:

- Prioritisation of waste issues or flows and the means to address them accompanied by assessments of the scale and costs involved.
- Targets for waste prevention or reduction.
- Identification of market failures and solutions.
- Partnerships with industry (producers, processors and retailers) to promote sustainable production and consumption, including improved infrastructure and technology and development of business models for waste reduction.
- Waste research and monitoring leading to advisory services on waste reduction, use and recycling.

- Capacity development to implement the approaches linked to awareness creation and information dissemination, including consumer awareness campaigns on food waste reduction.
- Support for innovation in waste technology and accompanying financial instruments.

Initiatives in OECD countries include establishment of robust and utilitarian waste definitions, standards and objectives that can be readily adopted in industry segments or processes and for waste data collection and aggregation; means to characterise more accurately the GHG impact of waste in fisheries and means to reduce the environmental footprint; incentives to encourage suitable wastes to be delivered to food charities (Box 2.9). Other programmes disseminate knowledge on the businesses case for the most high value waste reduction opportunities; assess the costs and benefits of using waste levies to offset local authority expenses and underpin the business case; and where technological gaps exist develop "innovation marketplaces" to foster solutions (see, for example, http://wbi.worldbank.org/developmentmarketplace/).

Backed by regulatory and traceability requirements, OECD processing and distribution chains tend to have relatively high efficiency and low waste. The perishable nature of fish products complicates marketing and logistics, and so is a key source of avoidable waste and losses, exacerbated by erratic capture fisheries supply and a lack of a cold chain where needed. Consequently, policies that increase market efficiency and help match supply with demand can reduce waste.

Classically, public support was provided for installation of ice machines at landing ports, boxes, wholesale and retail marketplaces, auction halls, product standards, price information and associated systems. Public support has extended to information sharing, efforts to stabilise markets and reduce transaction costs through cooperatives and producer organisations, and processor contracts and pricing formulas. For example, sales organisations in Norwegian fisheries are given public authority but are financed by the fishermen themselves through a fee on the first sale of fish.

Modern electronic auctions are highly effective in simultaneously addressing market efficiency, payment and logistics challenges and lessons learned from these can be applied to improve markets for waste products for secondary processing, conversion or disposal. Waste audits can be facilitated to assess and benchmark the level of avoidable waste produced by processing operations, such as filleting, shucking or picking to help processors exploit potential markets for wastes and thereby move the materials up the waste hierarchy. Protocols and standards for "marketable wastes" can be developed and waste volumes information synthesised to structure 'regional waste collection concessions' for processors in isolated areas.

Box 2.9. US regulations set a basis for reduction of food waste and hunger

Through Seashare (www.seashare.org) the seafood industry has donated 150 million seafood meals to the poor and hungry since 1994. Catalysed by the At-sea Processors Association, over 120 seafood companies and their associates have shipped seafood donations to 80 major food banks in 38 states, making them one of the largest sources of protein for hunger-relief in the United States.

The strategic partners include NOAA, national and regional food banks, financial contributors, service contributors such as cold storage firms, haulage (transport) companies, stevedoring and freight firms. The firms coordinate logistics through Seashare to collect, transport and deliver the surplus food.

In 1976, the US Congress enacted Section 170 of the Internal Revenue Code to encourage donations by allowing corporations to earn an enhanced tax deduction for donating selected surplus property, including food. The Code provides that wholesome food that is properly saved, donated to an approved agency and properly receipted is eligible for an enhanced tax deduction. This enhanced deduction is equal to half of the donated food's appreciated value, with the limitation that the total deduction cannot exceed twice the donated food's basis cost. This incremental tax deduction is calculated from the donated food's fair market value and basis food and labour cost. The authorities may challenge the value of donated food.

Source: www.seashare.org.

Improved allocations to maximise value

Fisheries management is as much about allocations of rights and access as it is overall stock management. The allocation problem is a complex one, coming as it does on top of a long history of traditional use that can be difficult to change. The following section on the dealing with the human dimension of reform will consider this in greater detail. That said, allocations of resources should be considered as part of a GGS in fisheries, as finding the highest-valued use of the resource is fundamental to delivering increased value from it.

Rights and responsibilities should be assigned as unambiguously as possible. Ambiguity leads to conflicts, bargaining and politicization. There are two types of rights to be taken into consideration: property rights (*à la* Coase-Alchian), and decision rights (*à la* Ostrom). Private property rights take many forms from the classical (as typified by ITQs) to collective property rights. This includes allocation of rights to a community over certain territories (TURFs), jurisdiction over the allocation of property rights allocated to national authorities through international treaties (RFMOs). As for decision rights, they are also different from property rights in that the latter concern the rights to use and benefit from a resource while the former concern the way actors (individuals or collective) operate with respect to property rights. Unfortunately there is no such a thing as a complete allocation of all rights and there will always be room for interpretation to test the robustness of institutions.

Allocations take place in a number of dimensions - within the commercial fisheries sphere and between commercial fishers and other users. This goes beyond sharing access to the fish resource, it can also be about access to marine areas, particularly coastal spaces that have many competing uses, from aquaculture to energy generation to shipping to recreation. Allocations can be improved through better planning in the early stages. In the European Union, new legislation will create a common framework for maritime spatial planning and integrated coastal management. While each EU country will be free to plan its own maritime activities, local, regional and national planning in shared seas would be made more compatible through a set of minimum common requirements.

The first impulse from the perspective of economics is to arrange sharing of the resource according the marginal benefit derived from it from each user group. This "equimarginal" approach forms the basis of neoclassical theory of resource allocation. In the case of fisheries, where allocations within groups may not be optimal (because of the absence of market transactions within groups), this logic does not always work. Abbot (2013) explains this in the context of recreational vs. commercial allocations (Box 2.10)

Resource management policies for commercial and recreational sectors, while fulfilling many other objectives, are unlikely to be designed to allocate access to harvest or fish mortality according to the rank ordering of willingness to pay. Only competitive markets, Pigouvian taxation and certain auction designs are likely to approach such efficient within-sector allocations (Abbot 2013). Holzer and McConnell (2013) note that the use of such mechanisms in resource and fisheries management is relatively rare.

It has been argued that sectors with the highest value should receive increased allocations. For example, estimates of the value of recreational fisheries that are higher than commercial fisheries are used as evidence that more fish should be made available to recreational fishers. But this confuses total and marginal values; it is unclear whether allocating more fish to the recreational sector in this case leads to higher social value (Edwards, 1991).

It therefore is difficult for the fisheries manager to define *ex ante* an optimal allocation of fish resources between all users, at least on economic grounds. This leaves two options: Allocations based on other criteria (historical amounts, social objectives, negotiation) and allocations using market-based measures such as tradable quotas. These are not mutually exclusive; it is possible (as one example) to have an allocation by group based on historical or other means that is decided administratively combined with trading within groups. In fact, this approach has been taken in many fisheries to obtain the efficiency benefits of tradable rights while still controlling consolidation and distribution of fleets to meet social objectives.

Box 2.10. The equimarginal principle and its shortcomings

Most discussions of efficient allocation across fishery sectors or across users within a sector make extensive use of the "equimarginal principle." The equimarginal principle begins by positing two sectors, each with their respective "derived demand" curves for fish mortality/harvest as an input. These demands reflect the horizontal summation of individual demands in each sector and obey the "law of demand" so that the marginal value of fish within a sector is declining in the number of fish allocated to that sector, the result being that the total value of increasing the allocation to a given sector increases at a decreasing rate (see figure below). In the case where an "interior" solution is efficient (so that it isn't efficient for some user groups to be deprived of allocation altogether) the equimarginal principle states that the efficient allocation occurs where the marginal benefit of an additional unit of allocation is equalised. This logic has been repeatedly expressed in the literature on discussions of recreational versus commercial allocation (Bishop and Samples, 1980; Easley and Prochaska 1987; Edwards 1991; Green 1994; Plummer, Morrison and Steiner 2012) and directly utilized in assessments of status-quo and prospective allocations of quota for policy making (Agar and Carter, 2012b; Agar, Carter and Waters, 2008; Gentner et al., 2010; Plummer, Morrison and Steiner 2012). Whenever the demand schedules for all sectors undergoing reallocation are known then it is a simple matter to solve for the efficient allocation. In the more common case, when only "local" information on demand for fish is known for one or more sectors, then comparing the respective marginal values at the current allocation between the recreational and commercial sectors provides qualitative guidance on the direction in which allocations should be adjusted but little guidance on the magnitude of the implied adjustment (Agar, Carter and Waters, 2008; Gentner et al., 2010).

These arguments, while having the intuitive appeal of simple textbook microeconomics, are predicated on a seemingly innocuous but restrictive assumption – that access to fish mortality within the sectors in question is actually allocated in an efficient manner – in a manner consistent with the ordering of marginal WTP underlying the derived demands of each sector. In other words, at a putative efficient allocation (see figure below), the system of property rights within each sector must offer exclusive access to fish mortality to each user in quantities up until the point where their individual demand intersects the "equilibrium" marginal benefit (the implied price in an inter-sector market) between the sectors. Users in either sector whose marginal valuation of the first increment of fish falls below this level must be excluded from the resource altogether.

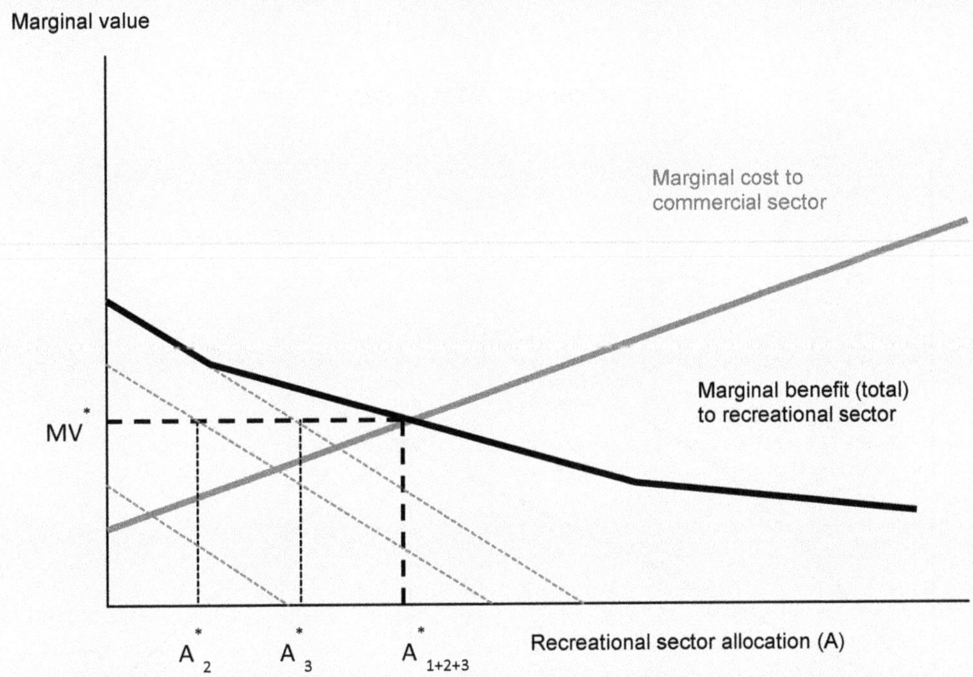

Efficient allocation of quota between sectors (A^*_{1+2+3}) and efficient allocation across three anglers in the recreational sector (A^*_2 and A^*_3). Notice that angler one receives nothing under this efficient allocation.

Source: Abbot (2013), *Integrating Recreational Fisheries into Fisheries Management: Challenges and Opportunities*, OECD internal document.

Harvest control rules, MSY and MEY

In order to manage a fish stock, the relationship between harvest and stock must be well understood. A harvest control rule identifies the harvest level that meets a defined objective for any given stock size. This begs the question: what rules, what objectives, and what timing? Recovering a depleted stock requires reducing the harvest below the growth rate at the current stock size. Recovering a stock faster requires aggressive reductions in harvest rates, which come at the price of greater short-term hardship for the participants in the fishery (see following section).

MSY, a biological target, has been the traditional choice for a targeted stock level, being the largest sustainable harvest rate. MSY is by definition the highest yield available from the resource and so provides the highest downstream benefits to consumers and processors. MEY, an economic target, incorporates stock effects on the cost of fishing effort in order to maximise the profitability of fishers. Higher stock sizes increase catchability and so lower cost for a given harvest level. Ideally, fisheries managers should take this into account when setting TACs, reducing the optimal harvest level to obtain a higher stock size than would be the case under MSY. However, fishers' true cost of production is private information that the manager has only imperfect information about. Moreover, the problem is more complicated when not all fishers are alike. The classical diagram showing MEY (Figure 2.7) assumes that all fishers are the same and have the same costs. When fishers are heterogeneous, the benefit of increasing the stock by reducing harvest depends on who ultimately catches the fish.

The increased information requirement and analytical complications that come from using MEY as the target of a harvest control rule make it practical in only a few situations. It makes sense only when an ITQ system is in place that equates fishers marginal cost of production and prevents cost-increasing competitive behaviour among fishers. In a TAC with open access (derby fishing), the race to fish that results leads to higher operational costs as well as higher capital costs as fishers invest in capacity to harvest rapidly.

Figure 2.7. MSY vs. MEY

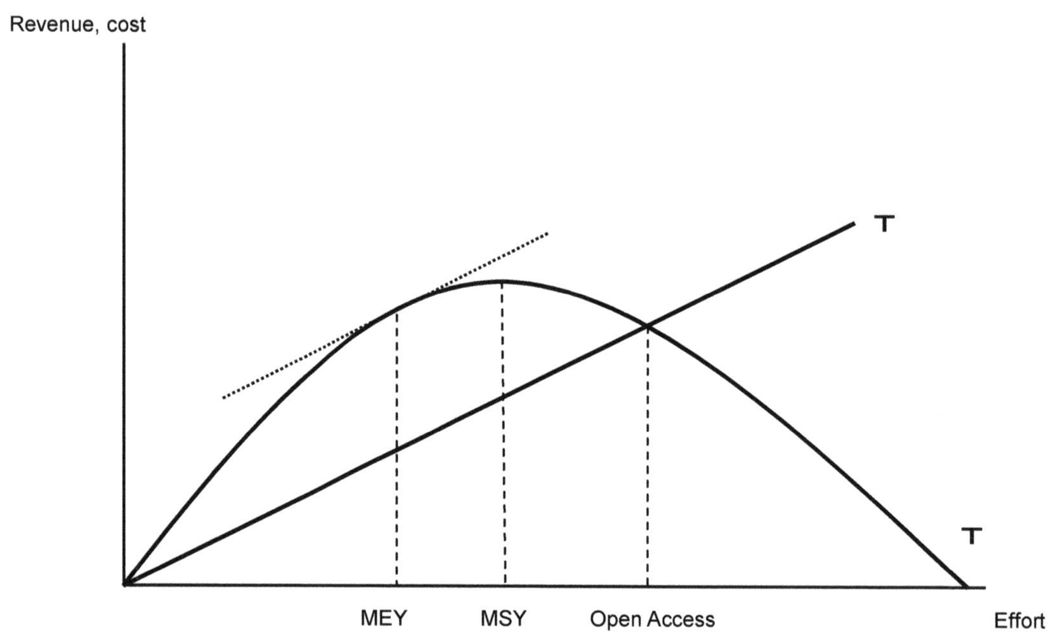

Source: OECD (2013a), *The OECD Handbook for Fisheries Managers: Principles and Practice for Policy Design*, OECD Publishing. doi: 10.1787/9789264191150-en.

MEY is often preferred by conservationist organisations as it is more conservative—it demands lower harvest and higher average stock sizes than MSY. The implication is that the stock is less likely to be overfished or suffer a collapse as a result of overfishing. It is generally true that managing stock as a higher percentage of carrying capacity reduces the probability of a stock collapse due to natural fluctuations, errors in stock measurement, or poor control of harvest. However, if these sources of uncertainty are real they should be reflected in the harvest control rule set by the fisheries manager directly, and not via the imperfect proxy that MEY provides. MEY is only coincidentally the right harvest control rule under uncertainty.

Enable change

Deal with human costs

Fishing is an important part of the coastal communities where it operates. It provides jobs in rural areas where often few alternatives exist. For many, it represents tradition and a way of life worth preserving (Box 2.11). As the fishing sector consolidates and modernises, it will be unable to play that role in every place where it has done so in the past. Modernisation has brought other challenges as well; fishing power and capacity have increased dramatically with new technologies. This has led to concerns regarding overcapacity and its negative implications for the biological and economic health of the sector where management has been inadequate.

Everyone agrees that fisheries should be sustainable and the resource stock should be well-managed in order to generate the greatest benefit for all. Despite this agreement, nations have struggled to put their fisheries on a sustainable footing and many continue to use their fish stocks inefficiently - depleted stocks, damaged ecosystems, and costly fishing and fisheries management make the stock generate much less value than it otherwise might.

At the same time, the cost of reform for many seems too high. It implies too great a change from tradition, means losses for incumbents, implies unknown risks and disruption and hardship during the transition period. If governments are to pursue a green growth agenda and enable change, they will have to find ways to mitigate and overcome these costs. Providing assistance with the costs and process of adjustment (flanking measures) can be effective, as can providing compensation to those who lose out from reform (OECD, 2007).

Changes in the fishing sector have been both promoted and resisted in OECD countries. Modernisation schemes provide funding to bring vessels into line with new technologies, while decommissioning schemes seek to reduce the number of active vessels. Research and development of new fishing technologies is pursued by governments directly and through aids to the private sector, while limits on vessel size and capacity try to reduce effective fishing power to prevent overfishing. In some places, tradable quotas have led to dramatic consolidation, while in others licenses and fishing opportunities are distributed in order to preserve existing fishing metiers and segments.

In each case, these actions are responding to the human dimension of the fishing sector. Those that promote modernisation and consolidation are focussing on the profitability of fishers, while those promoting diversity and tradition are focussing on the role of the fishing sector in the community. These are all valid choices, but in order to be part of a green growth strategy they must ultimately lead to greater efficiency in resource use and higher value-added generated by the sector. Durable progress in social objectives must be achieved from a solid economic foundation and not by fighting economic forces.

Box 2.11. Preserving small communities in the United Kingdom

Policy makers in the United Kingdom have stressed the desire to keep small peripheral communities alive. In fact, discussions often focus on the need to ensure that small fishing communities not only survive for the benefit of the fishing community itself, but also as part of broader objectives (e.g. ensuring fishing for tourism purposes or ensuring a local supply of marine protein). This has often led to a "ring fencing" of small-scale artisanal fisheries, endowed with a special status and rules in the fisheries management system. The following is an excerpt from a parliamentary commission discussing proposals for a transferrable rights system:

"The Common Fisheries Policy should protect fishing communities as well as fish. The introduction of Transferable Fishing Concessions (TFCs) as a mechanism to reduce fleet capacity highlights a broader debate over the interaction between overfishing, fleet size and employment in coastal areas. We recognise that introducing TFCs can reduce fleet capacity and improve environmental outcomes. However, we are deeply concerned that introducing TFCs will damage the viability of coastal communities. Department for Environment, Food and Rural Affairs (Defra) must decide what shape of fishing industry it wants in future. Therefore if Defra believes that a reduction in fleet capacity is needed, safeguards must be put in place to protect coastal communities and prevent excessive consolidation of the fleet in favour of larger operations.

"... In order to protect coastal communities from the potentially negative impact of fleet consolidation, Defra should not extend a system of Transferable Fishing Concessions into the under 10 m sector. Additional safeguards could include a limit on the percentage of national fishing concessions that can be held by a single vessel, a one-way valve to prevent transfers from small scale operations to large-scale operations, and a facility to allocate additional concessions to vessels that provide additional social or environmental benefits.

"...If a system of Transferable Fishing Concessions is introduced, Defra should implement a mechanism to discourage leasing of quota and to redirect unused quota towards more environmentally and socially sustainable fishing operators. We propose a siphon mechanism whereby if an operator chooses to lease his fishing rights rather than use them himself, a percentage of his allocation is returned to the national envelope. This can be reallocated to active fishermen in such a way as to restore traditional fishing activities in coastal communities and ensure the continuance of the socio-economic benefits that these activities provide."

Source: House of Commons Environment, Food and Rural Affairs Committee (2012), *Twelfth Report: EU proposals for reform of the Common Fisheries Policy*, HC 1563-I The Stationery Office Limited, UK.

Fleet capacity and adjustment

OECD countries have dedicated a relatively large share of budgetary support to fleet adjustment schemes, though this has declined in importance in recent years. Funding for these schemes has also come from industry or NGO contributions, and in many cases the total cost is shared by governments and others.

Decommissioning programmes have been demonstrated to be a useful policy tool, but only in certain circumstances. Done right, they can accelerate the elimination of overcapacity, but without associated reforms the progress they bring can prove temporary as capacity returns to the fishery. As part of a package of transitional assistance and management changes, they can provide a window of opportunity to help transform the nature of a fishery from one characterised by non-cooperative behaviour to one in which incentives are well-aligned and cooperation is the rational outcome of interactions between fishers (OECD, 2009).

The fact that decommissioning schemes tend to work well only when they are part of a larger reform package demonstrates the importance of an integrated approach as advocated by the OECD GGS. They are a tool for adjustment and compensation, not a management tool to control effort and are therefore no substitute for a well-functioning management system (OECD 2013). This view is reflected in the OECD Council Recommendation on Principles and Guidelines for Decommissioning Schemes (Box 2.12), which emphasises the role of the management system and the desirability of decommissioning schemes that are limited in time.

Box 2.12. Principles and guidelines for decommissioning schemes

Principles

Decommissioning schemes provide a useful mechanism for reducing capacity in situations where there is overcapacity. They can be used when urgent action is required to bring fishing capacity in line with available fisheries resources.

Taking preventative measures to avoid overcapacity from occurring is preferable to using decommissioning schemes to adjust capacity. Fisheries management systems should be appropriately designed to prevent overcapacity and overfishing from occurring, and to ensure that there are appropriate incentives for fishers to automatically adjust fishing capacity and effort.

The search for a perfect measure or a perfect assessment of capacity should not delay action to address overcapacity, although it is necessary to have an agreed measure of capacity to implement and enforce a cap on or reduction in capacity.

Decommissioning schemes should be designed to achieve the "best value for money", representing a cost-effective investment of public funds to achieve given capacity reduction objectives. They should be well-targeted and time-limited.

Decommissioning schemes will not, on their own, address the fundamental problems of overcapacity and overfishing. Decommissioning schemes should be designed as part of a package of adjustment measures towards sustainable and responsible fisheries. Social measures to assist retraining of fishers and community adjustment should be considered as part of fisheries adjustment packages.

Guidelines

Design

Decommissioning schemes should have well-defined objectives that are clearly articulated and measurable in order to ensure that the reduction targets are achievable and will have a positive impact on resource sustainability and economic profitability.

It is essential that the full range of management policies in place for the fishery, including the decommissioning scheme, are coherent and mutually supportive.

Governments should ensure that the management regime in place following the completion of the decommissioning scheme effectively prevents capacity from re-entering the target fishery or other fisheries, otherwise the beneficial effects of decommissioning will be negated over the medium to longer term.

Governments should ensure that the incentives of fishers are appropriately aligned in order to facilitate autonomous adjustment in the fishery in the future. This can be done by improving the specification and enforcement of access rights (based on either output or input dimensions) which will help to address the market failures that lead to the overcapacity problem.

Decommissioning schemes should be designed as part of one-off structural adjustment programs in order to avoid becoming incorporated into the expectations of the sector and distorting current and future investment incentives and plans.

The expected benefits and costs of decommissioning schemes should be evaluated during the design phase in order to ensure that the scheme will result in a net increase in economic welfare.

Governments should facilitate stakeholder involvement in the design and implementation of decommissioning schemes. This will improve acceptance of and compliance with the schemes' objectives and operations. The use of pilot programs may help. Stakeholder involvement will also improve the likelihood of cooperation in the post-adjustment management of fisheries.

Implementation

In implementing decommissioning schemes, governments should ensure that the criteria for determining the recipients of decommissioning pay-outs are transparent.

The mechanisms to determine the prices paid to decommission vessels, permits, licences and other entitlements should provide the best use of public funds in terms of impact on capacity and profitability. Where practical, governments should employ auctions to determine the prices and recipients of decommissioning pay-outs as this will generally provide the most cost effective means of determining prices and result in the most economically efficient allocation of resources.

Where more specific targeting of fleets or licence holders is required, other mechanisms such as fixed rate payments may be less complicated and costly to implement and should be considered by governments. Governments should ensure that such mechanisms are transparent and targeted, and that they minimise the transactions costs involved in their use.

Governments should target both latent and active capacity to ensure that capacity is effectively reduced and that capacity does not become reactivated in the fishery following the decommissioning scheme. Governments should take into account the potential impact of sequential decommissioning of latent and active capacity on resource sustainability and economic profitability.

Box 2.12. Principles and guidelines for decommissioning schemes (*cont.*)

Under the beneficiary pays principle, governments should require those who benefit from a decommissioning scheme to contribute to the costs of the scheme. A combination of industry and public funding improves the incentives for cooperative management of the fishery as the remaining fishers have a stronger stake in the future of the fishery, particularly if there is sound fisheries management in place.

Ex post evaluations of decommissioning schemes, linked to measurable performance indicators developed in conjunction with the scheme's objectives, should be undertaken to improve transparency and accountability. This will also help to ensure that the design and implementation of future schemes is informed by the experience of prior schemes.

Source: OECD (2009), *Reducing Fishing Capacity: Best Practices for Decommissioning Schemes*, OECD Publishing. DOI: 10.1787/9789264044418-en.

Political economy

Just because reforms are necessary does not make them easy, nor the choices obvious. There can be considerable agreement on the nature of existing problems and the solutions to them, and it can still be difficult to make progress. All reforms create winners and losers, and it is not possible to give everybody everything they want. Also, while the industry generally accepts that some fishing operations will fail due to normal business problems, it does not readily accept that failures be caused by fishery management actions (McLoughlin and Findlay, 2005). Stakeholder groups as a result may stall or prevent needed reforms or demand compensation for them. Fisheries managers need effective strategies to respond to public and stakeholder pressure in a way that builds consensus for reforms, meets government objectives and treats those affected in a fair and balanced way.

Economists generally advocate ITQs as a good way to fix the common-property problem in fisheries, but these systems are seldom welcomed by fishermen when they are first introduced. The reason for this may be that fishers have "non-transferable inframarginal skill rents" that they are unwilling to give up (Johnson and Libecap, 1982; Karpoff, 1987). That is, more skilled fishermen think they can do better than other fishermen under the current system, and want to preserve that advantage. Such fishers would prefer buyout schemes that eliminate their competition rather than ITQ systems (Reimer, Abbott, and Wilen, 2014).

Economic circumstances faced by the sector can narrow the set of politically feasible alternatives. It is more difficult to introduce reforms that temporarily reduce harvests when many fishers have low incomes and few alternative employment opportunities exist. A contributing factor to the collapse of the Atlantic Cod stocks in Canada was the regulator's concerns that reduced harvests would generate bankruptcies and unemployment. Moreover, when the sector is otherwise in good economic shape, a negative shock will have a less critical economic impact and the management response is usually more effective (Grafton et al., 2006).

Wherever policies have an impact on profits, stakeholder lobbying to influence the size and distribution of gains can be expected. Regulatory or budgetary policies can have an important effect on incomes, return on investment and other outcomes, so stakeholders will naturally seek to influence them. Lobbying activities aim to maximise benefits for the group doing the lobbying, which may not match the general interests of society as a whole. Part of the policy development and reform process consists in dealing constructively with lobbying activity to maintain stakeholder support without sacrificing broader social goals and benefits.

While the term "lobbying" can hold negative connotations, it is just a way for stakeholders to make a case for their interests. To ensure that policy decisions reflect the interests of all stakeholders in a balanced way, formal processes and contact points within institutions should be used. Without an official channel, lobbyists will seek to find the most influential ear or the most effective way to make their point. This could result in strikes, communication through the press, or seeking to make deals directly with politicians. These approaches put lobbying out of the control of the management institution and can lead to relationships with the sector that are less constructive.

Deal with the costs of reform faced by stakeholders

Objectives for mitigating human costs of reform while maximising value and conservation do not always mix easily. For example, the new EU CFP mandates reaching maximum sustainable yield for all stocks by 2015 where possible and the latest by 2020. This requires setting a harvest rule based on stock assessments that would lead to the MSY stock being reached by the proposed date. However, stock assessments can be variable, and the allowed harvest in a given year would vary with it. When new stock assessments lead to severe reductions in allowable harvest, dislocation and crisis can result (Box 2.13). Moreover, as pointed out in *Rebuilding Fisheries: The Way Forward* (OECD 2012), different paths for fisheries recovery may be appropriate depending on the conditions and consequences of each (Figure 2.8). Specifically, the speed of recovery depends on the rate of preference for harvests now vs. the future, the amount of harvest required to keep the sector viable during the recovery period and the degree of depletion of the stock and its risk of collapse.

Figure 2.8. Three possible recovery trajectories

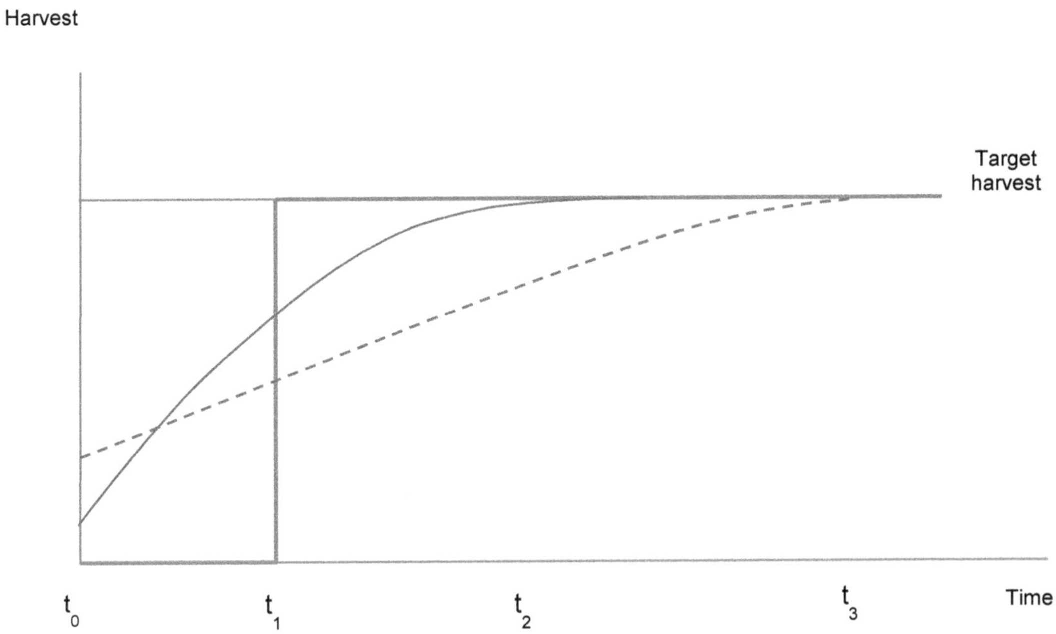

Source: OECD (2012), *Rebuilding Fisheries: The Way Forward*, OECD Publishing, Paris.
DOI: http://dx.doi.org/10.1787/9789264176935-en.

In terms of the economic and biological optimum, the idea of a constant-escapement policy as a best response to yearly variation has been known for some time (Reed 1979). This idea holds that all variation in stocks due to external forces should be taken into account immediately by adjusting harvest levels to keep the spawning biomass (the "escapement" after harvest) constant. As a practical matter, placing all annual variability into the harvest level is too disrupting for the sector and relies on very accurate stock assessments. As a result, variation is usually accommodated more slowly over time to balance variability in income and risks to stocks. How quickly to do this is fundamentally a problem of political economy; balancing the benefits of rapidly adjusting stocks to their optimal size against the dislocation caused by large or rapid changes in allowable harvest.

Limiting annual variation in harvest stabilises the sector at the cost of increased risk to the fish stock. An alternative way to deal with stock variation is to use flanking measures such as disaster payments to smooth fishers' income when the TAC is reduced. This is more disruptive to the operation of the industry (and upstream actors such as processors) but provides greater protection for the stock and a greater likelihood of stock recovery in a shorter period.

Box 2.13. Impact of quota reduction in the US Northeast

Quotas in the US Northeast groundfish fishery were sharply reduced in 2013 subsequent to revised stock assessments, including much lower quotas for some key groundfish stocks. In anticipation of these cuts, the Department of Commerce pre-emptively declared a fishery disaster in the fall of 2012, opening the possibility of federally-funded disaster assistance to the affected region.

"We know that for some fishing communities that have relied heavily on cod, haddock and flounder, the next several years are going to be a struggle," said John Bullard, NOAA Fisheries northeast regional administrator. "We've done everything we can to include measures that may help soften the blow of quota cuts, but it's going to take a collective effort to find more ways to keep both the fishery and the businesses that support it viable while these stocks recover."

NOAA Fisheries is taking a series of steps to help fishermen adjust to these measures, including:

- Implementing an increase in quota for healthier stocks such as redfish, white hake, and pollock. Knowing the challenges facing groundfish fishermen, NOAA Fisheries adjusted the 2013 white hake quota upward by about 15% over the proposed level, because recent analysis shows the stock condition has improved.
- Revising the rebuilding program for southern New England/Mid-Atlantic winter flounder. As a result, the catch limit for this stock will be increased by more than 150% over 2012, and generate an estimated USD 5.4 million in additional ex-vessel revenue for the fishery.
- Allowing some uncaught quota from last year to be carried forward into this year, reducing minimum legal sizes to allow more of the fish that are caught to be landed, and reducing some requirements for reporting, monitoring, and on small handgear operations.
- Allowing sector vessels to submit requests to NOAA Fisheries to fish in portions of areas that otherwise have been closed to fishing.

Source : NOAA press release, April 2013
http://www.nero.noaa.gov/mediacenter/2013/04/2013_groundfish_measures.html.

Managers have several ways of dealing with the human costs of reform. Following are five common approaches.

Reliance on existing social assistance: Easy, low-cost and administratively simple, this approach is best used for smaller reforms, but may not be politically feasible in small rural areas with longstanding economic problems. The additional social security available to fishers in France and Norway when direct fisheries supports are modified or withdrawn is one example of this approach.

- *Fiddling with reform*: Features longer phase-in periods, exemptions and carve-outs and other selective changes designed to enhance the acceptability and reduce the impact of reform. While this approach has the advantage of being targeted, its compromises can become permanent and sabotage adjustment. Hence, using it can damage credibility and raise doubts about the government's commitment to reform.
- *Economic diversification*: Aims at making the local economy and labour market more resilient to reforms. Active labour market programmes – e.g. employment insurance, early retirement, counselling and training, regional supports, aid to industry and infrastructure investments – have proven successful, are well understood and have a high level of political acceptability. However, regional support systems can preserve inefficient industries and distort markets, especially when alternative industries are already receiving support.
- *Compensation*: Effective at obtaining reform, but carries risks. Matching its level to stakeholders' perceived costs is difficult as estimates often do not include stakeholders' positive adjustment actions. Optimal compensation has more to do with the amount required to reach an agreement than with indemnifying losses; full compensation is not necessarily an objective. Compensation should always be targeted, time-limited and tailored to allow adjustment and minimise costs and market distortions.

- *Packaging reforms*: Combining changes in management with changes in supportive policies can ease adjustment costs and lead to greater efficiency and economic opportunity. New Zealand eliminated subsidies in the early 1990s at the same time as it introduced rights-based management. The new system gave those remaining in the fishery a good chance of creating a profitable business and allowed them to buy out leavers (OECD, 2007).

Ensure growth is inclusive

Much of the motivation for reform in fisheries has to do with improved efficiency—better resource use, more income for fishers and a greater economic contribution by the sector. But issues of fairness can dominate the debate when it comes to deciding the specifics of reform. Fairness in fisheries reform can mean many things, but in practice has meant preserving existing allocations, privileging incumbents over newcomers, and preventing fishers from being forced out of the sector.

Fisheries managers and policy makers need to work co-operatively with stakeholders to deliver effective outcomes. But providing the latter with incentives to participate must go hand in hand with a fair distribution of the benefits of reform to all. Finding this balance, and offsetting the influence of lobbying activities, requires broad consultation (primarily with fishers' organisations and secondarily with broader civil society groups) and strong connections among fisheries policy makers and other government agents. Bridge-building between thematic areas of government is already an important feature of policy coherence. Ideally, government ministries or agencies responsible for implicated domains – such as environment, natural resources, industry or employment – should have a formal mandate and role.

In fisheries, inclusiveness is often assured through fishers' organisations, regional councils and other local authorities to whom authority for allocation of fishing opportunities is delegated. Co-management can be understood as a modern version of community organisation that has developed as an alternative to more centralized forms of governance (Box 2.14). Although there are many different forms of co-management, they all depend on shared responsibilities defined more or less explicitly through negotiation and approval among concerned communities. In modern economies they almost always include public authorities as one party to the arrangement, who may have final authority in the case of disputes (Menard, 2014).

In almost every case, when fisheries reforms have instituted individualised rights to a fishery, the allocation of these rights has been done on the basis of historical participation in the fishery (Box 2.15). Rights allocated on this basis satisfy one of the primary considerations regarding fairness, by allowing those who have participated in the fishery in the past to continue to do so if they choose. Allocating these rights without charge is also typical. This serves to provide a benefit to rights holders that serves as an incentive to accept the associated reforms and often also reflects the situation where the value of the rights to a depleted fishery is small.

Box 2.14. Scallops in the Bay of Saint-Brieuc

An illuminating example of co-management involving fishermen, scientists and public authorities is provided by the exploitation of scallops in the Bay of Saint-Brieuc (France). At the beginning of a campaign, professional vessels collect samples to evaluate the biomass and make exploitation advices accordingly. Based on this scientific expertise, the shellfish commission of the Regional Fisheries Committee defines levels of capture and submits this proposal to public authorities for formal validation. The fishers are responsible for its implementation (time of fishing, duration, etc.) while the State services do part of the control (e.g. aerial surveillance). In case of infringement (excess collect, illegal trafficking), the representatives of the profession can file a complaint against the offenders and/or define internal penalties (e.g. suspension of licenses). Implemented in the 1980s, the system is considered a success: it has enabled fishers to reduce their operating costs, to significantly improve marketing of their product, and to reinforce the sustainability of their activity.

Source: Weigel, J.Y. and D. de Monbrison (2013), *State of the Art of Fisheries Co-Management*, Synthesis report to the Sub-regional Fisheries Commission, Agence Francaise de Développement, February.

Box 2.15. Initial allocations

A number of countries have proposed to auction fishing rights but have backed down owing to opposition from fishers who would have had to pay for a right that they currently enjoy free. This was the case in Estonia, for example (Eero et al., 2005, OECD, 2009). Fishermen considered it unfair to pay for fishing rights while competing in international markets with fishermen who did not pay for such rights. At the same time the implementation of the auctions themselves became problematic as the bidders engaged in cooperative behaviour in the bidding process, which is a well-known problem in auction theory (Laffont and Tirole, 1983). Similarly, in Russia the auctioning of quotas was introduced in 2001 but was for all practical purposes abandoned in 2004 (Honnelund, 2005). The only seemingly successful case where fishing rights have been auctioned is in the Washington State Puget Sound geoduck fishery, which has specific characteristics and is managed under a devolved management system with extensive stakeholder participation (Huppert, 2005).

Source: Haraldsson, G. and D. Carey (2011), "Ensuring a Sustainable and Efficient Fishery in Iceland", *OECD Economics Department Working Papers*, No. 891, OECD Publishing, Paris. http://dx.doi.org/10.1787/5kg566jfrpzr-en.

This is in contrast with the approach taken in the telecommunications sector, where rights to use frequencies are auctioned to the highest bidder, sometimes for billions of dollars. In principle, auctions are a superior approach from the perspective of inclusiveness, as the value of the rights accrues to the general public through the revenue raised. This is uncommon in fisheries because unlike in the case of allocations of spectrum, the resource has already been in legal use such that a *de facto* distribution of access already exists. Another factor is that participants in auctions for spectrum are large companies with sufficient capital and relatively secure business plans, while fishers may find financing a rights purchase difficult, and the benefits of doing so dependent on an uncertain outcome for the stock. Inclusive growth also means between fishers and the rest of society. In most countries, fishing represents a small share of GDP, so benefits-sharing is not a key consideration by policy makers who are free to maintain a sector-specific focus. But when the value of quota rights increases to an unexpected degree, the resulting distribution of benefits can lead to pressures for change.

In most cases, ITQ rights are allocated during a reform occurring in the context of a depleted fishery. Initially, the low productivity of the fishery means the value of the quota is low. Subsequent recovery leads to higher harvests and a steadily increasing quota value. This was seen in fisheries in Alaska and the west coast of Canada (OECD, 2010) and has also been the case in Iceland, where quota values now amount to around 2% of GDP (Box 2.16). The value generated by the improvement in the resource in these cases remains with quota holders.

The current system in Iceland is under pressure because quotas were initially freely allocated to fishers on a historical basis instead of being auctioned off or sold. Such methods for initial distributions are commonly used because the initial value of quotas are low for depleted fisheries (which lowers the stakes for negotiation), and they are likely to satisfy current fishers. This initial distribution may be perceived as unfair because the resource rent becomes concentrated in quota holders rather than broadly distributed. Calls for redistribution came after the fact of the initial allocation because of the steadily increasing value of the quota combined with the economic pressures on the wider Icelandic economy in the wake of the financial crisis of 2008. It has been proposed that this imbalance between fishers and the public be rectified by increasing the fisheries resource rent tax or by confiscating ITQs and auctioning them (Haraldsson and Carey, 2011).

The experience in Iceland and elsewhere indicates that plans for rights allocations in fisheries should include contingency planning for when the values of those rights increase significantly. Correcting imbalances after the fact should be avoided as this can add uncertainty, be unfair to rights holders who have purchased their rights after the initial allocation (thus shifting rather than solving the fairness issue), can hinder good management, and may be illegal in some cases (Haraldsson and Carey, 2011).

Building in advance some mechanisms for redistribution into the initial rights distribution can deal with fairness issues without introducing additional uncertainty, and there are a few ways this can be done. One way to do this is through pairing allocations with financial options based on the value of the fishing right allocated, such as by requiring holders to sell call options to the government. This is a way to cap the value of the transfer delivered to recipients of the initial allocation. By giving the government the right to buy back the allocations at some point in the future at a price fixed in advance, future increases in values beyond that fixed price can be more broadly shared.

By way of example consider an initial allocation of ITQs where the stock is depleted and the value of the allocation low, say EUR 10. The government buys a call option that is deeply out of the money, say EUR 100, paying the rights holders a fair market value for this. If the value of the ITQ remains below EUR 100, the call is worthless and the fisher keeps both their quota allocation and the payment received for the call option. If the price of the ITQ rises to, say, EUR 150, then the call is "in the money" - the government can buy the ITQ right for EUR 100, below its current trading value. The government can then redistribute this ITQ to new entrants, or sell it back to current fishers at its current price, netting EUR 50 which can then be redistributed as it sees fit.

More generally, if derivative markets are allowed to exist, the government could buy call options at any time and for any strike price. Because such markets would be voluntary, the fisher selling the option would be fairly compensated by the payment she receives for selling the call. In this manner, the government can cover most future scenarios regarding changes in value of the resource in a way that is fair to both current rights holders and others.

Another option is to define a resource tax that is a function of the value of the rights in advance. Such a tax could be designed to come into effect once valuations reach a certain multiple of their initial value. The beneficiaries of the option or tax could also be defined in advance, for example so that the local community could be certain of some benefit if the fishery increases significantly in value. For example, the resource tax could be defined as "10% of the value of rights over EUR 100 annually".

Box 2.16. Iceland's experience with ITQs

The ITQ system was introduced in 1984 for the cod fishery and subsequently applied to other species. With the Fisheries Act in 1990, all important fisheries were under an ITQ system. Under this system, each fishing entity owns or has a right to a certain percentage of the TAC in various species. These quotas are to a large extent tradable and can be leased with some limitations or sold. Small scale fishermen were originally excluded from the system and operated under effort limitations. Additional exceptions included measures such as regional quotas and special rules regarding long-liners.

Since the introduction of the ITQ system the industry has become much more economically efficient, with labour productivity now higher than in the Norwegian and Swedish fisheries (Eggert and Tveteras, 2007). The increase in efficiency has lifted the value of the resource rent and hence, of licences. Recent estimates of the net resource rent amount to ISK 14-34 billion (0.9-2.2% of GDP) per year (Kristofersson, 2010 and Steinsson, 2010)

Source: Haraldsson, G. and D. Carey (2011), "Ensuring a Sustainable and Efficient Fishery in Iceland", *OECD Economics Department Working Papers*, No. 891, OECD Publishing. http://dx.doi.org/10.1787/5kg566jfrpzr-en.

Measure progress towards objectives

Develop a framework and principles for measurement

Policies that promote green growth need to be founded on a good understanding of the different factors that affect green growth, and appropriate information is needed to monitor progress and measure results. Monitoring progress towards green growth requires indicators based on internationally comparable data. These need to be embedded in a conceptual framework and selected according to well specified criteria. Ultimately, they need to be capable of sending clear messages which speak to policy makers and the public at large. As part of its Green Growth Strategy, the OECD has developed a conceptual framework and indicators that help governments monitor progress towards green growth.

OECD green growth indicators are embedded in a conceptual framework which is structured around four groups to capture the main features of green growth.

- **Environmental and resource productivity**, to indicate whether economic growth is becoming greener with more efficient use of natural capital and to capture aspects of production which are rarely quantified in economic models and accounting frameworks.

- **The natural asset base**, to indicate the risks to growth from a declining natural asset base.
- **Environmental quality of life**, to indicate how environmental conditions affect the quality of life and wellbeing of people.
- **Economic opportunities and policy responses**, to indicate the effectiveness of policies in delivering green growth and describe the societal responses needed to secure business and employment opportunities. (OECD, 2011b)

Building indicators requires significant investment, and there are many practical difficulties related to defining and collecting the data needed. By working together, OECD countries can share these costs while having a consistent framework and approach. Many challenges still remain, such as:

- There are significant gaps in environmental-economic data at the sector level.
- There is a need to develop and improve the physical data for key stocks and flows of natural assets. Prominent examples are information on land and land use changes and non-energy mineral resources that often constitute critical inputs into production.
- Better physical data also helps improving material flow analyses.
- Improved information on biodiversity.
- Efforts should also be directed at developing monetary values reflecting prices and quantities for (changes in) key stocks and flows of natural assets. Such valuations, even if incomplete and imperfect are required for extended growth accounting models, more comprehensive balance sheets and for adjusted measures of real income.
- Periodic information to inform on how environmental concerns trigger innovation in companies should be developed.
- Thought should be given as to how indicators on economic instruments can be complemented by indicators on environmental regulation.
- Improved measures are needed on both the objective and the subjective dimensions of quality of life, in particular measures of environmentally (OECD, 2011b).

Deliver on green growth

The OECD Green Growth Strategy envisions an economy-wide approach to tackling the effects of the economic crisis while at the same time putting growth on a more sustainable footing. It encompasses a vast number of policy measures: fiscal reform; regulatory policy reform; changes to innovation policy; jobs strategies; climate change mitigation instruments; energy efficiency measures; competition policy in network industries and more (OECD 2011c). In this context, applying green growth to fisheries has two aspects. One is finding the role the sector can play in this economy-wide process of reform towards green growth. This means taking part in broader initiatives such as eliminating environmentally-harmful subsidies, preserving natural capital, and tackling climate change. The other aspect is applying the broader principles of the GGS to the sector-specific context of the fisheries industry.

In terms of the role of fisheries in the broader reform context, first priority must be seen as policy coherence - bringing fisheries policy planning and objectives more closely in line with national ones. After this comes policies related to climate change, which is a global issue against which every sector must do its part. Moreover, this is an issue that impacts fisheries directly: A warming, acidifying, rising ocean poses risks and adds unpredictability to the future prospects for fishing. Given this, the continued use of fuel tax concessions that encourage the use of fossil fuels, is counterproductive. Reforming these however is a particular challenge and perhaps the only way to accomplish this is in the context of an economy-wide review of environmentally-harmful subsidies. This is in line with the commitment by

G20 Ministers to "rationalize and phase out over the medium term inefficient fossil fuel subsidies that encourage wasteful consumption." (G20 Pittsburgh Declaration 2009) Finally, fishers have good reasons to want to ensure the sustainability of marine resources, but they are not the only ones. Preserving the richness of the marine ecosystem is an issue that goes far beyond the bounds of the fisheries sector. Properly valuing marine natural resources for purposes of decision-making must take a broader view than fishing, or even direct users, to include non-use valuation.

Learn from success and failure at home and abroad

This report has emphasised the importance of setting objectives and establishing accountability, and the virtues of measuring progress. In a well-designed institutional framework, this will bring about a positive cycle of continuous improvement via useful and actionable feedback as the situation in the fisheries evolves.

Such "at home" lessons of failure and success are important, but these lessons can also be learned from the shared experience of different countries. Adoption of ITQs accelerated after early successful implementations, and this is partly due to the demonstration effect provided by early adopters. As more examples of a system are put in place, it becomes possible to investigate their effects in a more rigorous way (Costello et al, 2008, Brinson and Thunberg 2013).

One of the core missions of the OECD is to bring member countries together to share experiences in a way that benefits everyone. The COFI's work on fisheries includes numerous workshops bringing stakeholders and governments together (OECD, 2010, 2012) and COFI participants have contributed a large number of case studies over the years (see for example OECD 2006, 2007, 2009, 2010a, 2010b, 2011).

Building an inventory of experience, and subjecting that where possible to scientific analysis, offers the potential to avoid the risk of mistakes, make the argument for reform, and reduce costs. Unfortunately, many of the well-run fisheries in the world have gone through at least one period of collapse or crisis, either economically or biologically. Investing in knowledge-sharing can help avoid these expensive lessons.

Box 2.17. What have we learned from attempts to introduce green-growth policies?

The OECD took stock of progress in implementing green growth policies since the release of the 2011 document *Towards Green Growth*. While there has been some progress in advancing the green growth agenda, this has been slow and much remains to be done. The main conclusions of this exercise were:

- Green-growth policies are likely to have beneficial welfare effects in the long term, but short-term transition costs have hampered their implementation.
- Despite some progress, green-growth frameworks remain limited in scope. The main challenge here is to coordinate policies and to develop indicators and instruments to monitor implementation progress.
- Pricing instruments have been widely used in green growth strategies, but have also been complemented by regulations or subsidies that can address market and information failures and are more politically acceptable.
- Countries need to pursue efforts to manage natural resources in a sustainable manner. This requires the development of indicators to properly value natural resources.
- Innovation is key to foster green growth and could be encouraged by a mix of policies within a coherent framework. Technology transfers have an important role to play, as long as trade and financial flows can circulate freely.
- Countries are concentrating more and more effort to invest in resilient infrastructure and adaptation policies, but additional public and private financing need to be mobilised.
- One important challenge is to overcome resistance to reforms and to find ways to compensate losers in a cost-effective way.

Source: OECD (2013), "What Have We Learned from Attempts to Introduce Green-Growth Policies?", *OECD Green Growth Papers*, No. 2013/02, OECD Publishing, Paris.
DOI: http://dx.doi.org/10.1787/5k486rchlnxx-en.

As the OECD increasingly turns its attention towards non-member economies, development assistance, and developing countries, our capacity to share knowledge and experience becomes more important. The benefits to global fisheries that result will benefit everyone.

In *Towards Green Growth*, Green growth is seen as "a strategic complement to existing priorities and areas for environmental and economic policy reform. Green growth strategies should target areas where there is clear beneficial overlap between environmental and economic policy and focus on finding cost-effective ways of attenuating environmental pressures, to begin the transition towards new patterns of growth that will avoid crossing critical environmental thresholds" (OECD, 2011c). Achieving this in practice is not always easy, and the lessons learned in the process will be a valuable aid to future initiatives (Box 2.17).

Establish new policy set

In terms of applying the GGS to the fisheries sector specifically, OECD work on fisheries is in many ways already aligned with the priorities and perspective of the Green Growth Strategy. This is because the major policy challenge in fisheries has long been to obtain maximum economic and social benefits from the fisheries resource and how to manage and sustain that resource for the long term. How best to accomplish this has been investigated with respect to a number of specific subject areas, from the management system to social adjustment, from energy efficiency to the effects of globalisation. From this work a few specific and central messages have emerged:

- **Good stock management is the foundation of fisheries policy**. A healthy stock that is harvested sustainably makes fishers more efficient and profitable, reduces social pressure and conflict, provides the most food for consumers and opportunity for fish processors, wholesalers and retailers. Poor stock management will eventually undermine or undo all other policy objectives for the fishery sector. This includes international cooperation in stock management.
- **Improve the efficiency of fishing**. Just as a healthy stock is the basis of a healthy sector, an economically efficient fishery is essential for maximising the benefits of fishing for all. Overcapacity, over-dependence on subsidies, and stressed coastal economies are all signs of an inefficient fishing sector. It is difficult to conceive of a sector that can be a constructive contributor to social objectives in the long term while at the same time being less efficient than is possible.
- **Have effective institutions**. Market-based and community-based approaches to fisheries management have a proven track record in improving both stock management and economic efficiency. There are many different paths to take in applying such approaches as evidenced by the diversity of fisheries management systems operating in OECD countries. Beyond management systems, institutions of governance that are open, transparent, responsive, flexible, and which give a central role to scientific advice are seen as necessary for a healthy and growing sector and cooperative and constructive stakeholders.
- **Push the boundaries of fisheries management outward.** A whole-of-government approach to broad policy setting should be taken, using clear and measurable benchmarks for success and treating reform as an opportunity to continually evaluate and improve policies. In line with advances in our knowledge of marine ecosystem interactions, the scale of management should expand from single-species to multiple species, food webs, and eventually the ocean economy.
- **Governments should embrace their role as facilitators of growth** by removing barriers to change, supporting innovation, and using the power of markets to help fishers maximise their profitability.

Applying the GGS to fisheries is also about taking an approach to reform that relies on following a number of important steps. These are: identifying the risks inherent in the *status quo* and the benefits of moving to a green growth model, removing the barriers to achieving that vision for green growth, dealing with the human dimensions of reform, measuring progress and finally establishing a new policy set. Combined, this is a comprehensive approach to reforms that is based on evidence, transparency and a focus on results.

This is not a process with a simple start and end point. It embraces the concept of adaptive management and institutional learning such that reform is a continuous process. In fisheries this is especially important as the resource base on which the sector depends has fundamental biological limits and fluctuates according to factors that are incompletely understood. If growth is to be continuous, reforms that increase efficiency and productivity must also be continuous.

In this view of green growth, the government is the facilitator of growth, not the source of it. The GGS demonstrates that much progress can be achieved by eliminating roadblocks on the path to green growth. Finding a governance model that frees the sector to innovate and grow while protecting the environment and delivering on social objectives is not always easy. There is no single solution that will work for all countries. But part of the process is making a break from the past and recognising when a new reality requires new policy approaches.

Notes

1. These sections draw upon work carried out by the OECD Committee for Fisheries under its 2013-14 programme of work, which focussed on the particular topics of waste, energy and governance.
2. William A. Knudson (2008) *The Environment, Energy, and the Tinbergen Rule Bulletin of Science Technology Society Online First,* doi:10.1177/0270467608325375. "The Tinbergen Rule states that for each and every policy target there must be at least one policy tool. If there are fewer tools than targets, then some policy goals will not be achieved. Further complicating the public policy environment are the facts that some policy tools affect more than one target, some tools help achieve more than one target, and others, while meeting one target, make meeting other targets more difficult. Also, some targets are more efficient than others. If policy makers are going to be able to effectively meet their environmental and energy goals, a series of policy tools need to be developed."
3. This view is supported by research done long ago by sociologists; see Merton's view that the only efficient social norms are those internalised by actors.

References

Agar, J.J., D.W. Carter, and J.R. Waters. (2008), "Economic Framework for Fishery Allocation Decisions with an Application to Gulf of Mexico Red Grouper," *NOAA Technical Memorandum NMFS-SEFSC-576*:1-100.

Brinson, A.E., E. Thunberg (2013), *The Economic Performance of U.S. Catch Share Programs, NOAA Technical Memorandum NMFS-F/SPO-133*, August.

Bishop, R.C., and K.C. Samples. (1980). "Sport and Commercial Fishing Conflicts - A Theoretical-Analysis," *Journal of Environmental Economics and Management* Vol. 7(3):220-33.

Carter, D.W., and C. Liese. (2012). "The Economic Value of Catching and Keeping or Releasing Saltwater Sport Fish in the Southeast USA," *North American Journal of Fisheries Management*, Vol. 32(4):613-25.

Crilly, R. and A. Esteban (2011), *Value Slipping Through the Net*, New Economics Foundation, London.

Crilly, R. and A. Esteban (2012), *No Catch Investment*, New Economics Foundation, London.

Dalskov, A.F.J. (2010), "Workshop on Fully Documented Fishery", *DTU Aqua-rapport* No. 221-2010, Technical University of Denmark.

Easley, J.E., and F.J. Prochaska. (1987), "Allocating Harvests between Competing Users in Fishery Management Decisions - Appropriate Economic-Measures for Valuation," *Marine Fisheries Review* Vol. 49, No.3, pp. 29-33.

Edwards, S.F. (1991), "A Critique of Three 'Economics' Arguments Commonly Used to Influence Fishery Allocations," *North American Journal of Fisheries Management,* Vol. 1, No 2, pp. 121-30.

Eero, M., M. Vetemaa and R. Hannesson (2005). "The Quota Auctions in Estonia and their Effect on the Trawler Fleet". *Marine Resource Economics*, Vol. 20.

Eggert, H. and Tveteras, R. (2007), *Productivity Development in Icelandic, Norwegian and Swedish Fisheries. Working Paper No. 31/07*, SNF-project no. 5175: "Fisheries management and development in fisheries: Comparative studies of Nordic Countries", Nordic Council.

European Commission (2012), *Blue Growth opportunities for marine and maritime sustainable growth*, COM/2012/0494/Final, http://eur-lex.europa.eu/legal-content/EN/TXT/?uri=CELEX:52012DC0494.

FAO (2011), *Global Food Losses and Food Waste*, Food and Agricultural Organization Publications, Rome.

FAO (2002), "Implementation of the International Plan of Action to Prevent, Deter and Eliminate Illegal, Unreported and Unregulated Fishing", *FAO Technical Guidelines for Responsible Fisheries*, No. 9. Food and Agricultural Organization Publications, Rome.

FAO (2001a), *International Plan of Action to prevent, deter and eliminate illegal, unreported and unregulated fishing*. Food and Agricultural Organization Publications, Rome.

FAO (2001b), *Reykjavik Declaration on Responsible Fisheries in the Marine Ecosystem*, Declaration made at the Reykjavik Conference on Responsible Fisheries in the Marine Ecosystem, Reykjavik, Iceland 1 to 4 October 2001.

FAO (1995), *Code of Conduct for Responsible Fisheries*, Food and Agricultural Organization , Rome, http://www.fao.org/docrep/005/v9878e/v9878e00.htm.

Fresh Start Project (2013), *Mandate for Reform: Securing the Right Deal for the UK, Equipping the EU for a Globalised World*, http://www.eufreshstart.org/downloads/mandateforreform.pdf.

Global Ocean Commission (2014) *From Decline to Recovery: A Rescue Package for the Global Ocean*, Global Ocean Commission Report 2014, Oxford , http://www.globaloceancommission.org/wp-content/uploads/GOC_Report_20_6.FINAL_.spreads.pdf

G20 (2009), *G20 Leaders Statement* Pittsburgh Summit 2009 https://www.g20.org/sites/default/files/g20_resources/library/Pittsburgh_Declaration_0.pdf.

Gentner, B., J. Kirkley, P.R. Hindsley, and S. Steinback. (2010), "Summer Flounder Allocation Analysis", *NOAA Technical Memorandum NMFS-F/SPO-111*.

Green, T.G. (1994), "Allocation between Commercial and Recreational Sectors in Stressed Marine Fisheries," *Society & Natural Resources*, Vol. 7(1):39-56.

Haraldsson, G. and D. Carey (2011), "Ensuring a Sustainable and Efficient Fishery in Iceland", *OECD Economics Department Working Papers*, No. 891, OECD Publishing, Paris. DOI: http://dx.doi.org/10.1787/5kg566jfrpzr-en.

Holzer, J., and K. McConnell (2013), "Allocation without Property Rights", *Journal of the Association of Environmental and Resource Economists*, Vol. 1, pp. 209-32.

Honnelund, G. (2005), "Fisheries Management in Post-Soviet Russia: Legislation, Principles, and Structure", *Ocean Development & International Law*, Vol. 36, No. 2.

House of Commons Environment, Food and Rural Affairs Committee (2012*), Twelfth Report: EU proposals for reform of the Common Fisheries Policy*, HC 1563-I The Stationery Office Limited, UK.

Huppert, D.D. (2005), "An overview of fishing rights", *Review of Fish Biology and Fisheries*, Vol. 15, pp. 201-215.

Johnson, R.N., and G.D. Libecap. (1982), "Contracting Problems and Regulation: The Case of the Fishery", *American Economic Review*, Vol. 72, pp. 1005–22.

Kaffine D. and C. Costello (2011) "Unitization of Spatially Connected Renewable Resources," *The B.E. Journal of Economic Analysis & Policy Impact*, Vol. 11, Article 15.

Karpoff, J.M. (1987), "Suboptimal Controls in Common Resource Management: The Case of the Fishery," *Journal of Political Economy*, Vol. 95, pp. 179–94.

Laffont, J.J., and J. Tirole (1983), *A Theory of Incentives in Procurement and Regulation*, MIT Press, Cambridge MA.

Larsen, T.A. and F. Asche (2011), "Contracts in the Salmon Aquaculture Industry: An Analysis of Norwegian Salmon Exports," *Marine Resource Economics*, Vol. 26, pp. 141–150.

Levhari, D and L. Mirman (1980), "The Great Fish War: An Example Using a Dynamic Cournot-Nash Solution", *The Bell Journal of Economics*, Vol. 11, pp. 322-334.

Macfadyen, G., T. Huntington, R. Cappell (2009), "Abandoned, lost or otherwise discarded fishing gear, UNEP Regional Seas Reports and Studies No. 185," *FAO Fisheries and Aquaculture, Technical Paper*, No. 523. UNEP/FAO, Rome.

Martini, R. (2012), "Fuel Tax Concessions in the Fisheries Sector", *OECD Food, Agriculture and Fisheries Papers*, No. 56, OECD Publishing, Paris. DOI: http://dx.doi.org/10.1787/5k9bdccqft30-en.

McLoughlin, R. and V. Findlay (2005), "Implementation of Effective Fisheries Management", *Outlook '05*, Canberra.

Menard, C. (2014), “Institutional Aspects of Governance in Fisheries Management”, OECD internal document.

NOAA (2013), press release, http://www.nero.noaa.gov/mediacenter/2013/04/2013_groundfish_measures.html, April.

NRDC (2013), *Bringing Back the Fish, An Evaluation of U.S. Fisheries Rebuilding Under the Magnuson-Stevens Fishery Conservation and Management Act*, Report # R-13-01-A, February.

Plummer, M., W. Morrison, and E. Steiner (2012), "Allocation of Fishery Harvests under the Magnuson-Stevens Fishery Conservation and Management Act: Principles and Practice," *NOAA Technical Memorandum NMFS-NWFSC-115*:1-99.

Ten Brink, P. et al. (2009), *Guidelines on the Use of Market-based Instruments to Address the Problem of Marine Litter*, Institute for European Environmental Policy (IEEP), Brussels, Belgium, and Sheavly Consultants, Virginia Beach, Virginia.

OECD (2014), “Food Waste Along the Food Chain”, *OECD Food, Agriculture and Fisheries Papers*, No. 71, OECD Publishing, Paris.
DOI: http://dx.doi.org/10.1787/5jxrcmftzj36-en.

OECD (2013a), *The OECD Handbook for Fisheries Managers: Principles and Practice for Policy Design*, OECD Publishing, Paris.
doi: 10.1787/9789264191150-en.

OECD (2013b), “What Have We Learned from Attempts to Introduce Green-Growth Policies?” *OECD Green Growth Papers*, No. 2013/02, OECD Publishing, Paris.
DOI: http://dx.doi.org/10.1787/5k486rchlnxx-en

OECD (2013c), *Policy Instruments to Support Green Growth in Agriculture*, OECD Green Growth Studies, OECD Publishing, Paris.
DOI: http://dx.doi.org/10.1787/9789264203525-en.

OECD (2013d), *OECD Review of Fisheries: Policies and Summary Statistics 2013*, OECD Publishing, Paris.
DOI: http://dx.doi.org/10.1787/rev_fish-2013-en

OECD (2012), *Rebuilding Fisheries: The Way Forward*, OECD Publishing, Paris.
DOI: http://dx.doi.org/10.1787/9789264176935-en

OECD (2011a), *Fisheries Policy Reform: National Experiences*, OECD Publishing, Paris.
DOI: http://dx.doi.org/10.1787/9789264096813-en.

OECD (2011b),*Towards Green Growth: Monitoring Progress: OECD Indicators*, OECD Green Growth Studies, OECD Publishing.
doi: 10.1787/9789264111356-en.

OECD (2011c), *Towards Green Growth*, OECD Green Growth Studies, OECD Publishing, Paris.
DOI: http://dx.doi.org/10.1787/9789264111318-en.

OECD (2010a), *Advancing the Aquaculture Agenda: Workshop Proceedings*, OECD Publishing, Paris.
doi: http://dx.10.1787/9789264088726-en.

OECD (2011), *The Economics of Adapting Fisheries to Climate Change*, OECD Publishing, Paris.
doi: http://dx.10.1787/9789264090415-en.

OECD (2010c), *Interim Report of the Green Growth Strategy: Implementing our commitment for a sustainable future: Meeting of the OECD Council at Ministerial Level*, OECD internal document, Paris, 27-28 May 2010.

OECD (2009), *Reducing Fishing Capacity: Best Practices for Decommissioning Schemes*, OECD Publishing, Paris.
doi: http://dx.0.1787/9789264044418-en.

OECD (2008), *Recommendation of the Council on Resource Productivity*, OECD Internal document, Paris.

OECD (2007), *Structural Change in Fisheries: Dealing with the Human Dimension*, OECD Publishing, Paris.
doi: http://dx.10.1787/9789264037960-en.

OECD (2006), *Fishing for Coherence: Proceedings of the Workshop on Policy Coherence for Development in Fisheries*, The Development Dimension, OECD Publishing, Paris.
doi: http://dx.10.1787/9789264025301-en.

OECD (2002), "OECD Workshop on Waste Prevention: Toward Performance Indicators 8-10 October," ENV/EPOC/WGWPR/SE(2002)1/FINAL, OECD, Paris, http://www.oecd.org/officialdocuments/publicdisplaydocumentpdf/?doclanguage=en&cote=env/epoc/wgwpr/se(2004)1/final.

OECD (2001), *Extended Producer Responsibility: A Guidance Manual for Governments*, OECD Publishing, Paris.
DOI: http://dx.doi.org/10.1787/9789264189867-en.

Reed, W. (1979), "Optimal Escapement Levels in Stochastic and Deterministic Harvesting Models," *Journal of Environmental Economics and Management*, Vol. 6, pp. 350-363.

Reimer, M., J. Abbott, and J. Wilen (2014), "Unravelling the Multiple Margins of Rent Generation from Individual Transferable Quotas", *Land Economics*, Vol. 90, pp. 538-559.

Shelton, P. (2007), "The weakening role of science in the management of groundfish off the east coast of Canada", *ICES Journal of Marine Science*, Vol. 64. pp 723-729.

Skall, H.F. and N.J. Olesen (2011), *Treatment of wastewater from fish slaughterhouses*, National Veterinary Institute, Technical University of Denmark, June.

Sumaila, U.R., W. Cheung, A. Dyck, K. Gueye, L. Huang et al. (2012), "Benefits of Rebuilding Global Marine Fisheries Outweigh Costs," *PLoS ONE*, Vol. 7(7): e40542.
doi: http://dx.doi.org/10.1371/journal.pone.0040542.

Sumeth, P. (2010) "Contract Farming and Riverine Aquaculture," *Asian Labour Update,* Issue 74, January - March.

Sutinen, J. and R. Johnston (2003), "Angling Management Organisations: integrating the recreational sector into fisheries management", *Marine Policy*, Vol. 27, pp. 471-487.

Weigel, J.Y. and D. de Monbrison (2013) *State of the Art of Fisheries Co-Management.* Synthesis report to the Sub-regional Fisheries Commission, Agence Francaise de Développement, February.

World Bank (2013), *Fish to 2030: Prospects for Fisheries and Aquaculture*, World Bank Report N° 83177-GLB, Washington D.C.

Chapter 3

Green growth in aquaculture

This chapter identifies the challenges for green growth in aquaculture, the policies that can underpin further sustainable growth in aquaculture, and the factors necessary for successful aquaculture development. The objective is to develop advice and best practices that can inform a roadmap for national aquaculture planning. This report also discusses the effects on competitiveness of incorporating green growth principles into aquaculture policy.

This chapter identifies the challenges for green growth in aquaculture, the policies that can be brought to bear to underpin further sustainable growth in aquaculture, and the factors necessary for successful aquaculture development. The objective is to develop advice and best practices that can inform a roadmap for national aquaculture planning.

Aquaculture: A history of rapid growth, with concerns about sustainability

Aquaculture is the farming of aquatic organisms in inland and marine waters, involving intervention in the rearing process to enhance production and the individual or corporate ownership of the stock being cultivated (FAO, 2008). A wide range of species are farmed including fish, crustaceans and molluscs, bivalves and aquatic plants (seaweed). FAO recorded 310 species in 2008 as being cultured, excluding aquatic plants. In 2011, OECD countries produced about 6.85 million tonnes including aquatic plants of which diadromous fish (e.g. salmon) contributed 36.6% followed by molluscs (27.8%), aquatic plants (19.8%), marine fishes (8.8%), freshwater fishes (4.2%), and crustaceans (2.5%) (Source: Fisheries and Aquaculture Information and Statistics Service, FAO) (Table 3.1).

Aquaculture production has grown considerably over the past decades. FAO statistics suggest a production increase from 3.5 million tonnes in 1970 to close to 79 million tonnes in 2010 (Table 3.2). The *OECD/FAO Agriculture Outlook 2013* expects world fisheries production to expand to a total of 181 million tonnes by 2022 (which includes 15,6 million tonnes of fish for reduction to meal and oil), of which 85 million tonnes will come from aquaculture. These projections seem to be rather conservative compared to the increases that have taken place over the past two decades. Of the total aquaculture production of 78,9 million tonnes in 2010, 72,2 million tonnes, corresponding to 91.5%, was produced in Asia, predominantly by developing economies in the region.

Over the past four decades, the largest production increases have come from freshwater fish, seaweeds, crustaceans and molluscs. Production of marine fish (most of which are carnivorous) has grown more slowly and now makes up a relatively small portion of the total (Figure 3.1). The slower increase in the production of carnivorous species is likely due to the costs of feed and feed compounds which have increased markedly due to scare resource base (fish meal and oil). A key future challenge in this regard is innovations that allow substituting fish meal and oil with terrestrially-produced substitutes. Another factor for the relatively modest production growth for carnivorous species is the relatively high level of externalities in production and the resulting administrative limitations to growth.

Some emerging economies such as Viet Nam are important export-oriented producers of fish from aquaculture, while others, like China, mainly supply the domestic market. This diversity, together with poor reporting from many aquaculture-producing economies, makes it challenging to identify common features of green aquaculture. Ultimately, incorporating green growth principles in aquaculture will mean more efficient regulation of externalities and a better understanding of local impacts, but what this means in specific cases will depend on individual production systems as well as local and regional factors and practices. Meanwhile, a number of framework conditions for sound aquaculture development will be outlined in this study.

While aquaculture production has increased substantially, there are concerns about the sustainability of aquaculture production due to environmental externalities, supply of feed resources, and competition for space. For example, many shrimp aquaculture farms in Southeast Asia were set up at the cost of mangrove destruction, and later many of them were abandoned because of contamination (Allison, 2011). Escaped fish or disease transfer from aquaculture to wild population is also a concern (Bostock et al., 2010).

Table 3.1. Major aquaculture species in OECD countries

Species		AUS	BEL	CAN	CHL	CZE	DNK	EST	FIN	FRA	DEU	GRC	HUN	ISL	IRL	ISR	ITA	JPN	KOR	MEX	NLD	NZL	NOR	POL	PRT	SVK	ESP	SWE	TUR	GBR	USA
Aquatic plants	Japanese kelp																	x	x												
	Green seaweeds				x													x	x												
	Red seaweeds				x										x			x	x												
Crustaceans	Red swamp crawfish																x														x
	White leg shrimp																		x	x											x
Diadromous fishes	River eels	x					x	x			x	x					x	x	x						x		x	x			
	Atlantic salmon	x		x	x		x			x		x		x	x								x				x			x	x
	Rainbow trout	x	x		x	x	x	x	x	x	x	x	x	x	x	x	x	x	x	x	x		x	x	x	x	x	x	x	x	x
	Sturgeons, paddlefishes						x	x		x	x		x			x	x							x			x				x
Freshwater fishes	Grass carp					x							x			x								x		x					
	Silver carp		x			x							x			x	x									x					
	Channel catfish																			x											x
	Miscellaneous freshwater fishes	x	x	x		x	x	x	x	x	x	x	x		x	x	x	x			x			x		x	x	x		x	
	Tilapias and other cichlids															x	x		x	x	x									x	x
Marine fishes	Atlantic cod													x									x							x	
	Flounders, halibuts, soles																	x	x	x			x		x						
	Turbot				x		x			x				x							x				x		x			x	
	European seabass									x		x				x	x								x		x		x	x	
	Gilthead seabream									x		x				x	x								x		x		x		
	Miscellaneous pelagic fishes																	x	x	x					x						x
	Atlantic bluefin tuna											x					x								x		x				
Molluscs	Abalones, winkels, conchs	x			x					x					x				x			x					x				x
	Japanese carpet shell			x						x					x		x		x								x			x	x
	Mussels			x	x		x			x	x	x		x	x		x		x	x	x	x	x		x		x	x	x	x	x
	Pacific cupped oyster	x		x	x					x	x				x		x	x	x	x		x			x		x			x	x

Source: FAO (2014), *Fishery and Aquaculture Statistics. Global production by production source 1950-2012* (FishstatJ).
In: FAO Fisheries and Aquaculture Department [online or CD-ROM]. Rome. Updated 2014. http://www.fao.org/fishery/statistics/software/fishstatj/en

Table 3.2. Aquaculture volume and value, 1970-2010

Volume '000 tonnes, value USD millions

	1970	1980	1990		2000		2010	
	Volume	Volume	Volume	Value	Volume	Value	Volume	Value
Brackish water	185	368	1 305	5 271	2 124	9 001	5 222	15 440
Fresh water	1 270	2 343	7 631	11 951	18 477	22 111	37 017	69 409
Salt water	2 071	4 636	7 903	9 450	21 124	19 962	36 705	40 374
Total	**3 526**	**7 347**	**16 840**	**26 272**	**41 724**	**51 073**	**78 943**	**125 224**
Of which…								
Freshwater fishes	1 168	2 092	7 140		17 585		33 742	
Marine fishes	51	187	318		977		1 834	
Aquatic plants	959	2 641	3 765		9 306		19 007	
Crustaceans	10	87	755		1 691		5 725	
Molluscs	1 068	1 837	3 609		9 757		14 158	

Source: Food and Agriculture Organization of the United Nations (2012), *FIGIS*, Food and Agriculture Organization Publications, Rome.

Figure 3.1. Aquaculture production by type

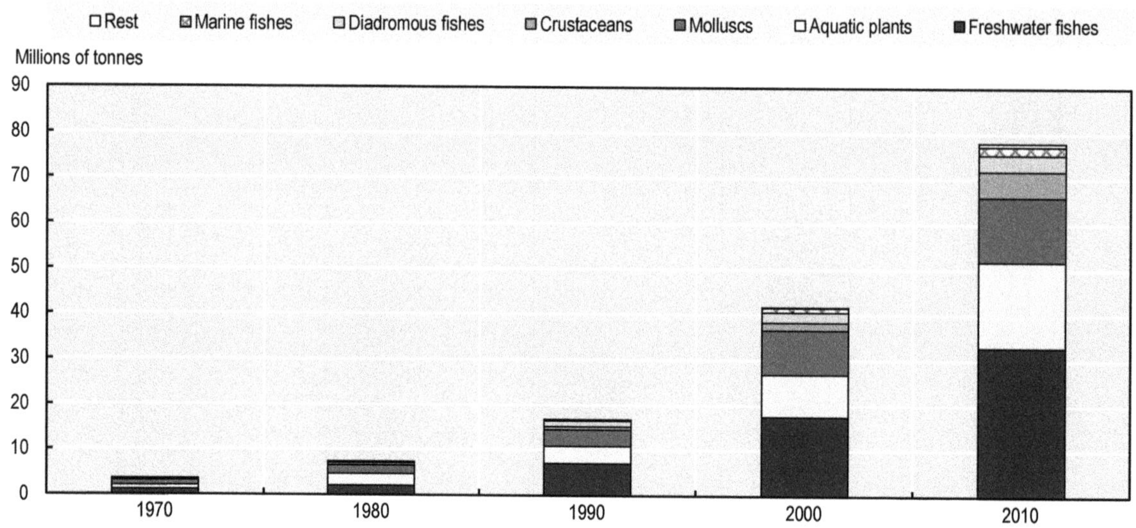

Source: Food and Agriculture Organization of the United Nations (2012), *FIGIS*, FAO Publications, Rome.

Concerns regarding the negative externalities of aquaculture have made it difficult for many developed countries to establish fish farming as a growth sector. After the rapid growth of the 1980s-1990s, aquaculture production stagnated in Europe and North America in particular mainly due to regulatory restrictions on site and inputs (Bostock et al., 2010). With a few exceptions, there has been no meaningful growth in aquaculture production in OECD economies, which accounted for 35% of the value and 30% of the volume of world aquaculture production in 1984, but only 18% and 9%, respectively, in 2007 (Figure 3.2).

On the other hand the same concerns have also encouraged the development of better practices and technologies which has resulted in more sustainable aquaculture practices in many OECD countries. For example, technology developments have improved monitoring and site selection tools, and led to new feeds that use less fish-based inputs, leave less emissions, and the use of vaccines instead of antibiotics. Seen in this light the restrictions or regulatory requirements on aquaculture growth have spurred innovations to address the concerns caused by production externalities. In this regard, an additional challenge is to change consumer's perception on aquaculture, i.e. the image of the aquaculture sector.

Figure 3.2. Total aquaculture production: Volume and value

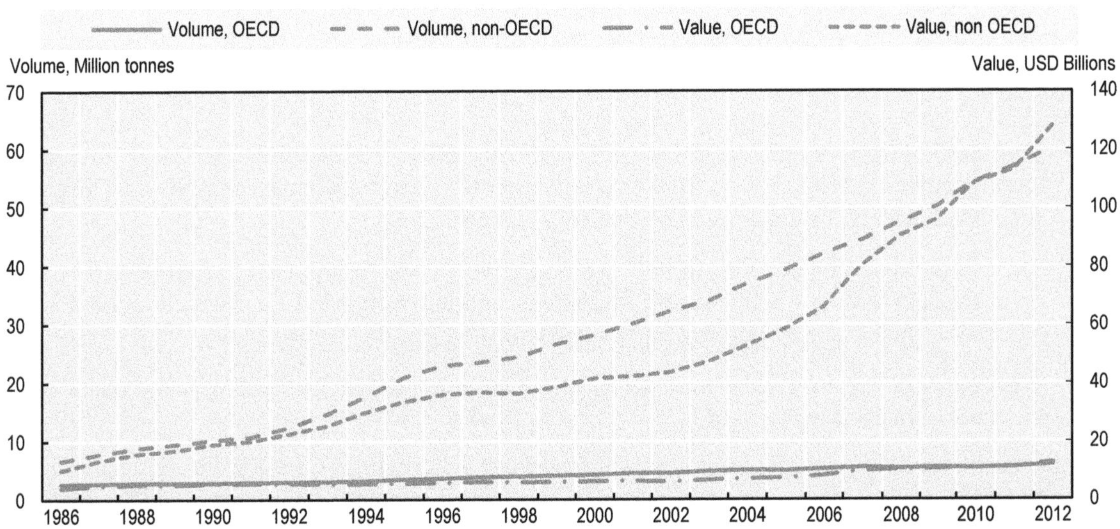

OECD (2014), "Fisheries: Production from aquaculture", *OECD Agriculture Statistics* (database). DOI: http://dx.doi.org/10.1787/data-00225-en.

Build a green growth model in aquaculture

The OECD COFI workshop on *Advancing the Aquaculture Agenda – Policies to Ensure a Sustainable Aquaculture Sector*, held in April 2010 in Paris, concluded that aquaculture has a high potential to contribute to economic growth and food security because good management practices make it possible to limit environmental harmful effects while increasing food production. Indeed, compared to the rearing of terrestrial animals, aquaculture offers significant advantages (OECD, 2010a).

Growth in aquaculture production in OECD member economies has been slow over the past decades (*Advancing the Aquaculture Agenda*) (Figure 3.2). Among countries that have similar conditions for aquaculture development, some have developed aquaculture while others have not. There may be, however, common features at play which may have created differences between the OECD member economies and non-member economies, as well as among certain OECD member economies. These features may be related to governance, technologies, environmental regulations or resource availability, e.g. space and natural endowment more generally. At the same time, there has been a significant increase in aquaculture production in Southeast Asian countries, including Viet Nam, Cambodia, Thailand, Myanmar, Indonesia, Malaysia and Philippines, since the mid-1970s. Government interventions, such as a stable licence scheme, the provision of seed, and financial incentives, are factors that have contributed to this growth, in addition to growing global market demand.

Major points raised by the presentations at the Green Growth and Aquaculture Workshop held in Yeosu, Korea (December 2012) included ensuring biosecurity, the use and potential of aquatic genetic resources; improving governance; consistency with international standards and interaction with other economic activities. The discussions summarised and highlighted three central themes: research, international co-operation and involvement of all stakeholders. Other important points highlighted included spatial planning, certification, and a need for institutional innovation. The participants agreed that green growth in aquaculture is feasible and desirable and an on-going process to be applied in aquaculture development frameworks and strategies, most notably in national development.

Address the risks posed by externalities in aquaculture

Aquaculture must continue to address the wide range of externalities it causes if it is to grow sustainably. As aquaculture competes for space with other users, both recreational and commercial, the path towards green growth must include spatial planning (including user conflicts), sanitary issues, licence systems, site allocation, and, importantly, cooperation among the various stakeholders in aquaculture, e.g. farmers, consumers, authorities (Table 3.3).

A green growth strategy for aquaculture would seek to continue the expansion of the sector while reducing its negative impacts and demand on natural resources. Aquaculture externalities are a function of:

- The species being grown (e.g. seaweed is mostly about competition for space; mussels have positive externalities as they consume excess marine nutrients; carnivorous species such as salmon or shrimps give rise to nitrogen discharges);
- The method used by the fish farmer (i.e. intensive vs. extensive, cage, recirculation, ponds or open marine farming) with lowest discharges coming from land based recirculation systems albeit such systems are relative energy intensive;
- The degree of human intervention for example in terms of disease fighting using medicines and the amount of feed being used.

Aquaculture policies should be tailored to address externalities specific to the species being produced and the production system; for example, the external impacts of carnivorous fish production in recirculated production system can be more easily contained than in open marine cage culture.

Table 3.3. Green Growth challenges and aquaculture

Green growth challenges	Variables to control	Policy framework	Measures (examples)
Feed resources	Feed	Innovation, fisheries management	Use grains and vegetables, use of wastes
Discharges	Feed, feed conversion ratio, feed components, water quality, siting of operations	Regulations, innovation, good management practices	Feed quotas, fallowing, cleaning, transferable discharging permits, taxes, IMTA, reuse, zoning, environmental monitoring, monitoring feeding
Diseases	Density, temperature	Regulations, innovation, good management practices	Distance, vaccine, limiting use of antibiotics, fallowing, zoning, environmental monitoring, screening protocol, crop calendar
Escapees	Storms, accidents, genetics, predators	Regulations, good management practices, Innovation	Stronger cages, sterilisation, paying local fishermen to catch escapees, use of native strains, genetics impact models
Space	User conflicts / conflicting uses	Coastal zone/ocean management, regulations	Reserved areas(zoning), spatial planning
Food safety	Toxic, drugs or environmental waste in product	Regulations, good management practices, enforcement capacity	Establishment of pre-approved zones for aquaculture development, enforcement, sampling and certification system
Regional development	Development planning	Permits and zoning, environmental approvals, Investment aids, coastal zone/ ocean management	Establishment of pre-approved zones for aquaculture development

Table 3.3. Green Growth challenges and aquaculture (*cont.*)

GDP contribution	Growth of sector, marketing of product	Marketing and promotion, research and development, infrastructure investments	Support private certification schemes, streamlining permit process
Development	Capital, skills	Education and training, labour standards	Continuing education for local populations
Energy use	Feed, electricity	Regulations, innovation, good management practices	Use grains and vegetables, Use of wastes, taxes, combine with wind mill, technology transfer to industry
Public perceptions and acceptance	Informed messaging of science-based information	Outreach and education	Working with press and NGOs, sharing information on websites, certification, labelling.

Source: OECD (2011a), *Towards Green Growth: Green Growth Strategy Synthesis Report*, OECD internal document, Paris.

The best response to these externalities depends on the direction of the externalities caused by aquaculture production:

- When the externality impacts the farmer himself (e.g. local water pollution harming productivity) it is self-interests that drives the response.
- When impacting other fish farms (e.g. disease outbreaks) in which case an industry association combined with public regulation might be able to address the problem.
- When impacting other sectors of the economy or the environment (e.g. aesthetic conflicts with tourism) in which case public regulation may be necessary.
- The challenge for public regulators is to find the right level of intervention (Box 3.1). The industry itself has an interest in addressing externalities as the negative effects may reduce profitability and demand in the long run (Asche, 2011) Certification schemes like the Global Aquaculture Alliance's "Best Aquaculture Practices Standards" or the Aquaculture Stewardship Council's certification program can help aquaculture producers adopt best practices and market their product on the basis that they are sustainably produced. Certification schemes can influence consumer concerns about aquaculture and thus help improve the image of the sector. The challenge for policy makers is to ensure that the economic contribution of the sector can continue to grow while minimising its negative environmental impacts. Regulations have many roles to play in meeting this challenge. They should stimulate investment, foster economic growth, reduce systemic risk, lead to environmental improvements as well as health and overall quality of life. The policy tools addressing externalities can be grouped in two categories, market-based instruments and non-market instruments (Table 3.4).

Box 3.1. Optimal regulation in aquaculture

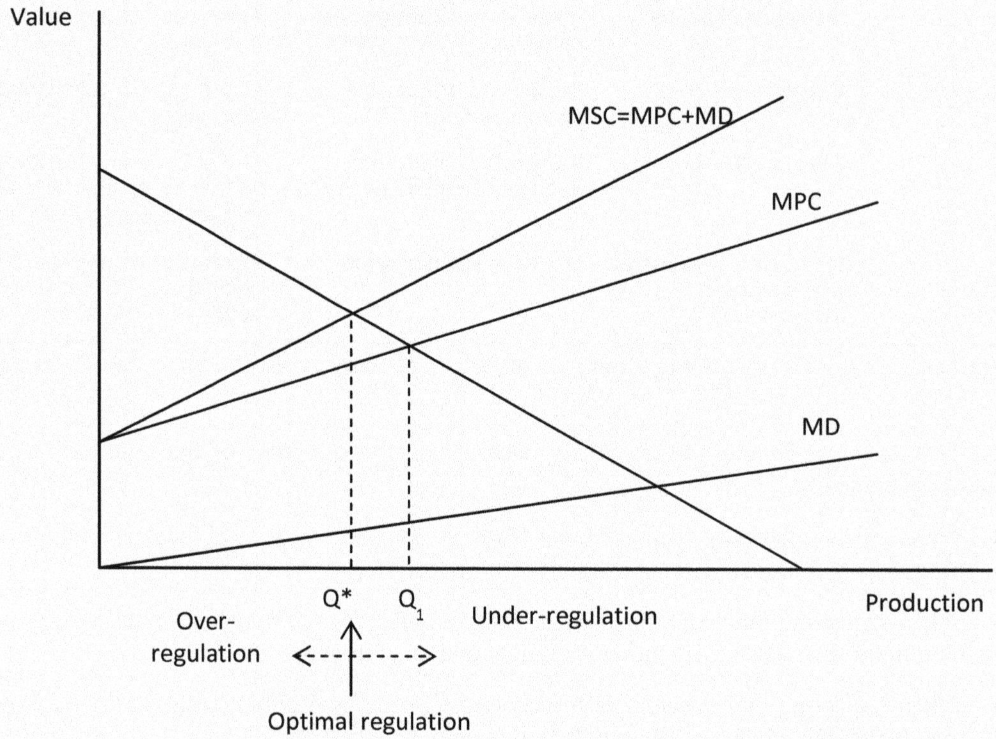

The relationship between the marginal private costs (MPC), the cost of producing an additional unit, and the level of production is shown in the figure. Marginal benefit (MB) is achieved from each additional unit produced. Marginal damage (MD) is damage done by pollution coming from production of each extra unit. The slope of the MD curve is positive indicating that each additional unit of pollution does more damage than the preceding. The marginal social cost (MSC) is the vertical summation of MPC and MD. Hence, MSC represents the total cost to society of the existence of production, including production costs and cost of damage from production done to other parties in society.

Given an unregulated market, the optimal production level and the corresponding optimal pollution level would be at Q_1, since this is where marginal benefits equal marginal production costs. However, in a situation where pollution is regulated, damage done to the environment is taken into account in identifying the optimal production level. In this case the optimal level of production (Q^*) is where the marginal social costs equal marginal benefits. Regulations that allow production above Q*, are too loose, allowing marginal social costs to exceed benefit. Regulations that lead to production below Q*, are tighter than necessary, implying that marginal benefit exceed social costs.

Source: Nielsen, R. (2012), *Barriers to Sustainable Growth in the Aquaculture Sector: An Economic Analysis*, Ph.D. Thesis, University of Copenhagen, Denmark.

Table 3.4. Policy tools to address externalities

	Market-based instruments		
	Strengths	**Weaknesses**	**Conditions for favourable use**
Cap-and-trade permit systems	Tend to equalisation of abatement costs	High start-up costs	Public-good market failure is not dominated by monitoring and information costs.
	Continuous incentives to innovation	Steep learning curve	Sufficient institutional capacity (experience) and sufficient size of market
	Certainty over emission levels	Lower adoption incentives due to costs	Damage depends on overall amount of a pollutant not on specific location or timing of emission sources
	Can raise revenue	Potential price volatility	Precise control over emissions is available at reasonable cost
	Once in place will be defended by stakeholders	Competitiveness and income distribution issue	Cross-border spill-over effects are important
Taxes or charges on pollution or exploitation of natural resource	Tend to equalisation of abatement costs	Potentially high monitoring costs	Public-good market failure is not dominated by monitoring and information costs.
	Can raise revenue	Lower adoption incentives due to costs	Pollution sources are small and diffuse
	Continuous incentives to innovation	Uncertainty about emission level	Environmental damage depends on overall amount of a pollutant and not on specific location or timing of emission sources
	Implementation through existing institutions	Lower adoption incentives due to costs	Temporary deviations in emission levels have little damage on environment
		Competitiveness and income distribution issue	Precise control over emissions is available at reasonable cost
Taxes or charges on a proxy (input or output)	Implementation through existing institutions	Loss of static and dynamic efficiency relative to charges at source	Control of direct pollution discharge difficult or costly
	Lower monitoring costs than permits or direct taxes		Close and stable relationship between use of input or output as proxy and targeted pollutant
			Several pollutants associated with single input or output
Subsidies	High adoption and compliance incentives than permits/ taxes	Potentially large budgetary costs	Enforcement of alternative pricing instruments is difficult or very costly
		May trap excessive resources	Activity to be subsidised is a strong substitute for targeted "dirty" activity
		Uncertainty about emission level	Subsidy programme can be designed in a relatively simple way, for a time-limited period and with minimal secondary effects
Command and control Performance standards	Leave flexibility to search for cheapest option	Do not naturally tend to equalise marginal costs	Pollution control at the source of emissions is infeasible or very costly
	Higher adoption & compliance incentives than pricing measures	Potentially high administrative costs	No adequate proxy for pollutant that could be object of taxation
	Certainty on emission level	More information required than for permits and taxes	Weak response of agents to price signals
			Pollution emissions can be measured from application of technology

Table 3.4. Policy tools to address externalities (*cont.*)

	Market-based instruments		
	Strengths	**Weaknesses**	**Conditions for favourable use**
Command and control Technology standards	Higher adoption and compliance incentives relative than pricing measures Low monitoring costs Certainty on emission levels	No flexibility to search for cheaper abatement costs Not easily adaptable to new costs and benefits info. No incentives to innovate	Pollution control at the source of emissions is infeasible or very costly No adequate proxy for pollutant that could be object of taxation Administrative costs of performance standards are too high Abatement costs are relatively homogeneous across agents
Active technology support policies	High adoption and compliance incentives High incentives to invest in R&D of new technologies	Do not directly address externalities Potentially large costs/ Uncertainty on emission levels	Technology areas where market size and learning-by-doing effects are dominant Infrastructures in areas where network considerations are important
Voluntary approaches	Contribute to information gathering and dissemination on costs and benefits High adoption incentives	No intrinsic mechanism to encourage adoption of least-cost options Uncertainty on outcomes, Risk of collusion among participants	When the authorities can put strong pressures (credible threat of follow-up actions) Where information is not too costly to provide

Source: de Serres, A., F. Murtin and G. Nicoletti (2010), "A Framework for Assessing Green Growth Policies", *OECD Economics Department Working Papers*, No. 774, OECD Publishing, Paris. DOI: http://dx.doi.org/10.1787/5kmfj2xvcmkf-en.

Market-based instruments aim to address externalities and promote green growth mainly through price signals designed to equalise the private and social costs and benefits of economic activities undertaken by private agents. These include taxes, charges and fees, tradable permits, and subsidies. Non-market instruments aim at addressing externalities through means other than price signals. They include direct environmental regulations (e.g. technology standards, performance standards and ban of certain activities), active technology support policies and voluntary approaches including information-based instruments.

In principle, market-based instruments provide agents with incentives and flexibility to search for the least cost ways of meeting environmental targets and thus are more cost-effective than non-market instruments. Non-market instruments generally do not meet cost-effectiveness criterion because they provide no intrinsic mechanism for meeting environmental targets at the least economic cost. However, there are cases where non-market instruments will be more cost-effective or appropriate than market-based instruments. These include:

- when monitoring and enforcement costs and information problems are large;
- when emissions cannot be observed or easily monitored and there are no appropriate proxies for emissions;
- when relevant resources or environments to be affected are too important or sensitive to allow temporary deviation from policy targets; and
- when information is lacking and costly to obtain, which leads to a weak response to price signals.

The best choice of instruments will differ across countries, environmental areas, region-specific circumstances, the nature and size of the predominant market failures and institutional capacities of respective countries, etc. In addition, a combination of instruments will usually be more appropriate given the presence of several interacting market failures. The following table summarises the main features of policy tools for externalities.

Discharges

Aquaculture activities interact with the surrounding environment. As aquaculture continues to intensify and expand, discharges of organic wastes, nitrogen and phosphorous may result in local environmental degradation. This is a main source of criticism of aquaculture and explains why strict restrictions on aquaculture expansion are in place in many countries. The use of antibiotics has also been a concern because of the potential harm to humans and the environment when dissipated into the surrounding water and taken up by other aquatic species.

Some proportion of feed provided to farmed species is diffused to surrounding water columns or accumulated on the bottom. The faeces from farmed fish are also diffused to water columns or accumulated on the bottom. In total, this can release more nutrients than can be assimilated by the surrounding environment. As a result, poor water quality, eutrophication or dead zones may appear. Eutrophication may lead to reduced dissolved oxygen and hypoxic or dead zones which may result in fish kills, excessive phytoplankton and macroalgal growth. The latter can reduce light penetration and be harmful to submerged aquatic vegetation or induce harmful algal blooms which may result in mass fish kills and decrease biodiversity due to changes in nutrient composition (Selman et al., 2008).

In order to manage discharges from aquaculture operations, many countries have implemented command and control measures such as maximum permissible discharge loads. For example, Denmark introduced strict environmental regulations to control water pollution from aquaculture activities based on a maximum allowable feeding, statistical standard for nitrogen, phosphorus and organic matter, a minimum level of oxygen in the outlet water, and a limit on water intake (Jarlbæk and Børrensen, 2012). Incentive-based policies such as transferable discharge permits, taxes and subsidies are another approach that offers flexibility and increased incentives for innovation. In February 2012, the Danish Ministry of Environment enacted a new regulation which introduces the option for fish farmers to voluntarily choose regulation on nitrogen discharges instead of feed quotas.

Fallowing, cleaning and integrated multi-trophic aquaculture such as combined farming of fish, filter feeders (mussels or sea cucumber, for example) with aquatic plants (like algae and kelp) are also possible options. Seaweed and molluscs not only remove discharges, nutrients and particulates from the water column including from aquaculture activities but also transform a substantial portion of them into their tissues (Box 3.2).

In general, production of aquatic animals leads to lower emissions of nutrients than terrestrial animals except for chicken. Bivalves extract nitrogen and phosphorus from the water column (Table 3.5) and can even be used to mitigate or remedy environmental degradation.

In different settings phosphorous and sludge from aquaculture production can be a valuable resource and capturing and re-using the resource is a new area for technology and regulatory development (Nordic Council, 2012). For example it may be a useful input in composting.

Choosing appropriate sites for production is often the key to reducing or eliminating impacts from discharges. In many cases just moving operations to deeper water with stronger currents can significantly reduce impacts as has been the case in Turkey (Box 3.3). Modelling carrying capacity to inform siting of operations can be of great assistance.

The important message is that there are solutions even if some of them may be expensive or operationally challenging. Fully re-circulated systems, for example, can eliminate local discharges and escapees although they still have discharges that need disposal (typically on land). And extensive production systems can recycle nutrients using multi-trophic aquaculture. Finally, identifying a maximum load on the environment for each kind of discharge and making emission permits transferable between farms and among industries also will allow for a more efficient allocation of resources.

Box 3.2. Aquaculture of filter-feeding molluscs

Freshwater fishes, aquatic plants and molluscs together have accounted for more than 80% of the world aquaculture production since 1970. Shellfish aquaculture has also increased considerably over the same period and accounted for about 94% of the world molluscs produced in 2010. Especially, shellfish aquaculture has drawn special attention from academia and policy makers due to the environmental services as well as social and economic services it provides.

In general, shellfish do not need to be fed in the rearing process because they are primary consumers unlike other aquaculture species. Shellfish feed on and filter nitrogen and other organic and inorganic materials in the water column. In doing so, shellfish contributes to improved water quality and habitat and stock restoration by reducing eutrophication symptoms and enhancing light penetration and providing oxygenation into water column in addition to providing social and economic benefits including food, job creation and poverty alleviation.

Ferreira *et al.* (2009) simulated the value of European shellfish aquaculture with five EU farms (ones in Scotland, France, Slovenia, Italy and Portugal) farming four major species (blue and Mediterranean mussels, Pacific oyster, and Manila clam). According to the study, the European shellfish farms remove 55 000 tonnes of nitrogen per year, which is equivalent to the nitrogen load of 17 million people. A study on mussel farming on the Swedish west coast (Lindahl et al., 2005) shows that mussel farming can be an innovative and strategic environmental management tool to recycle nutrients from sea to land. In this region, a hectare of mussel farm with vertical suspenders attached to horizontal longlines may consume the equivalent of 20 hectares of the annual phytoplankton production and produce about 300 tonnes of mussels in 12-18 months (Lindahl and Kollberg, 2009). Also, a kilogram of fresh mussels in this region are reported to assimilate 8.5~12 g of nitrogen, 0.6~0.8 g of phosphorous, and about 40~50 g of carbon (Lutz, 1980; Peterson and Loo, 2004). The mussels containing the excess nutrients can be used as sea food for human consumption or fertilisers on farmland or feedstuff for livestock. The value of removing nutrients by mussel farming was estimated by simple comparison of abatement costs with other abatement measures (Lindahl, 2005) and by the cost savings by the replacement of other abatement measures such as improved agricultural operations, improved sewage treatment and use of wetlands (Gren et al., 2009). Though the estimated marginal costs varied considerably depending on the growth rate of the mussels and the market for the mussels produced, mussel farming had lower marginal costs (cost savings between 2-11%) than many other abatement measures in the region (Gren et al, 2009). A rural person is reported to discharge 3.5-4 kg of nitrogen per year (Anonymous, 2003), so 350-400 kg of mussels per year is the production of mussels needed to abate it. A compensation scheme for mussel farmers for the nitrogen discharge can be devised based on this information.

From the integrated multi-trophic aquaculture (IMTA) cases shellfish farming can provide additional benefits compared to the monoculture case. A Farm Aquaculture Resource Management (FARM) model results for a 3.2 ha oyster farm cultured with a finfish farm in Sanggou Bay, northeast China show that nitrogen removal becomes seven times more and oyster production becomes thirty times more compared to the oyster monoculture case (Ferreira et al., 2011).

Notwithstanding these benefits, shellfish aquaculture has been blamed for some negative externalities such as eutrophication from bio-deposition on the sediments or damaging sediments. Burkholder and Shumway (2011) reviewed 62 ecosystems and found only four ecosystems with sustained system-level impacts while the other 58 ecosystems have either negligible or localised significant impacts from bivalve shellfish framings. Those negative externalities mostly occur when in high density culture in a poorly flushed area or when harvested mechanically.

Overall, the negative effects from shellfish farming are local phenomena in exceptional cases or negligible compared to the overwhelming land-based nutrients loading and the huge benefits from shellfish aquaculture (Burkholder and Shumway, 2011).

Source: Shumway, S.E. (Ed.) (2011), *Shellfish Aquaculture and the Environment,* Wiley-Blackwell, Singapore.

Table 3.5. Comparison of sustainability indicators for animal protein production system

	Feed conversion (kg feed/ kg edible weight	Protein efficiency (%)	N emissions (kg/tonne protein produced	P emissions (kg/tonne protein produced)	Land (tonnes edible product/ha)	Consumptive freshwater use (m3/tonne)
Beef	31.7	5	1 200	180	0.24-0.37	15 497
Pork	4.2	25	800	40	1.0-1.20	3 918
Chicken	10.7	13	300	120	0.83-1.10	4 856
Finfish (average)	2.3	30	360	48	0.15-3.70	5 000
Bivalves	Not fed	Not fed	-27	-29	0.28-20.00	0

Source: Phillips, Beverige and Clarke 1991; FAO 2003; Hall et al. 2011; Bouman et al., 2013 (recited from Notes June 2013, The World Bank).

Box 3.3. Resolving site conflicts: The case of Turkey

Aquaculture in Turkey has developed fast and has come under great pressure due to pollution as well as its unsightly appearance and unpleasant smell which has caused conflict with tourism. Cage farms that were entirely concentrated near the coast until 2006 were subject to major criticisms.

In January 2007, the Ministry of Environment issued a communiqué which stipulated that net cage farms shall be sited in locations which have at least a 30-meter deep water column, are at least 0.6 nautical miles away from land, and have at least 0.1 m/s of current speed. The same communiqué classified the sites where these farms can be located based on eutrophication. All existing farms that were located in near shore locations in Turkey have since been moved to locations which meet these criteria. During this period, three provinces where aquaculture was common were identified as potential aquaculture areas and farms were moved to these locations.

Since these farms were moved to new locations, short and mid-term problems have been resolved. However, given the high level of investment needed to implement these changes, many farms without sufficient capital were taken over by larger companies.

Escapees

The environmental effect of escapees is an important challenge in aquaculture especially in sea-cage farming. Interaction between wild and farmed fish may pollute genetic pools and reduce the survival capability of wild species. The escaped species may also compete with wild stocks for feed or become the dominant species, impacting biodiversity. If farmed fish are not indigenous to the area of production the escaped fish becomes an invasive species that may disturb the ecosystem. The escapees may also spread diseases or pathogens to the wild stocks.

Reliable and complete escapee data are not available on a global scale. However, Norway has collected comprehensive data on escapees. Fredheim et al. (2010) state there are over 325 million Atlantic salmon held in sea-cages in Norway, which is far greater than the wild salmon population of about 1 million. Since the escapee rate is a small share of total production (0.1-0.3%) it may not be a sufficient incentive for farmers to actively prevent escapements. However, there may be an indirect cost to the industry and society as escapees undermine the industry's reputation and can be detrimental to ecosystems (Fredheim et al., 2010). As part of the regulatory reforms that have been developed in recent years in Chile progress has been noted in both preventive and mitigation measures. In particular, escapees have been dealt with through the regulation of the security of farming structures.

Based on the data from sea-cage salmon farming in Norway the causes of the escapees can be broadly categorised into structural equipment failure (68% of escaped fish), operational related-failure (8%), biological (17%) and external factors (8%) which are also species dependent (Jensen et al., 2010). Though the structural failures are not frequent they tend to lead to incidents with a large number of escapees. In contrast, operational failures usually lead to small incidents but are more frequent. It clear that structural failures need to be addressed in order to prevent escapees (Fredheim et al., 2010).

Norway introduced in 2004 a technical standard for marine fish farms including regulations for design, dimensioning, production, installation and operation (Fredheim et al., 2010). These regulations were further strengthened in 2011. In addition, the Norwegian government has imposed an upper limit on the number of fish to be kept in each net pen. Combined, these two measures effectively reduced the overall risk of escapees both in terms of the numbers and as a proportion of number of fish in sea-cages.

There are some policy lessons to be learned from the Norwegian experience (Jensen et al., 2010 and Fredheim et al., 2010) including:

- Establish mandatory reporting system of all escapees.
- Establish a mechanism to collect, analyse and learn from the mandatory reporting.
- Conduct mandatory technical assessments on the cause of large-scale escape incidents.
- Introduce a technical standard for sea-cage aquaculture equipment.

- Conduct mandatory training of fish farm staff.
- Pay local fishermen to catch the escapees.
- Conduct R&D for better equipment, sterilisation and understanding species behaviour.

The degree of genetic separation between the wild and farmed populations is key to mitigating potential genetic impacts, as is the size of the wild population. Modelling such potential impacts can help aquaculture managers select proper strategies, as is currently done in the United States and the European Union.

Diseases and parasites

Aquaculture activities may transfer non-native and native diseases and parasites to surrounding production areas (e.g. ISA introduced in Chile) and to wild species through various ways such as eggs and fingerling transactions, equipment, fish-to-fish contact or currents. This often leads to a decrease in production and may sometimes carry significant economic losses and pose a threat to wild fish populations.

Asche et al. (2010) argue that disease is always present in any animal husbandry industry so disease control should be an essential part of animal farming including in aquaculture. Indeed disease outbreak was one of the most serious concerns among delegates to the "Green Growth and Aquaculture Workshop" Yeosu, Korea. The Chilean case can happen elsewhere; rapid expansion mainly driven by short-term economic interest and in the absence of appropriate regulatory framework leads to acute problems (Box 3.4). Good governance is very important in controlling disease. Addressing individual cases only when problems appear via allopathic measures such as heavy reliance on the use of antibiotics is less effective than good precautionary measures (OECD, 2010).

Precautionary measures include spatial planning to allow for periodical fallowing and relocation of farming sites, regulations to keep a certain distance among farms, limits on stocking density, vaccination of smolts and reducing the use of antibiotics. Such measures will all help to limit disease outbreaks.

The Chilean ISA outbreak in 2007 inflicted major economic losses on the sector and job loss in both farming and processing. the Chilean Government worked in close co-operation with industry to address the disease outbreak, demonstrating the value of a long term collaborative approach to aquaculture. Following the ISA crisis, Chile has been innovative in the design and implementation of precautionary measures (Box 3.4).

The 2004 case of white spot disease outbreak in Sri Lanka also caused huge economic losses to the tiger shrimp aquaculture industry (Box 3.5). While most Asian countries which experienced the same disease outbreaks had changed to *Penaeus vannamei* farming to prevent a reoccurrence, Sri Lankan producers continued with tiger shrimp farming and took various measures, such as shrimp farming zoning and the introduction of a crop calendar with the farmers' active involvement, to revive the shrimp farming industry.

Box 3.4. The recovery of the Chilean salmon industry

Over the last three decades the Chilean salmon industry has been impressively successful both in technical and commercial terms. Today, Chile is the second largest producer in the world. However, regulatory framework did not properly address biological risks and other social, economic and environmental issues. The industry's priority was on production, sales and overall economic benefits from aquaculture growth.

When the infectious salmon anaemia (ISA) virus outbreak occurred in 2007 the lack of a good regulatory framework impaired the industry's ability to respond. Production was hard hit, decreasing by 67% (from 376 476 tonnes in 2006 to 123 233 tonnes in 2010), accompanied by a significant fall in the number of Atlantic salmon farms In operation (from 375 in 2007 to 66 in 2009) and a 50% loss of direct and indirect jobs (around 25 000 lay-offs).

Some of the major reasons for the outbreak included the high concentration of farms within a limited area, the absence of zone management programmes, poor sanitary control on farms, high stoking numbers on farms, and a lack of transparency in the industry.

With a rapid public-private co-ordinated effort, basic infectious disease control measures were implemented to increase biosecurity on farms and included a quality assurance of diagnostic laboratories and mandatory reporting. At the same time, collaboration between the government, financial sector and industry was developed to finance the industry so that it could continue to operate. New laws and regulations were implemented to facilitate the industry's recovery in the long term.

Measures that have since been adopted include spatial planning to allow periodical fallowing and relocation of farming sites, regulations on distance between farms, limit on stocking density, vaccination of smolts, reducing use of antibiotics and modification of the regulation of import of eggs to make it consistent with international standards of the OIE (World Organisation for Animal Health) and to raise the level of sanitary protection in the country.

In 2011, the production volume began to increase and the stocking of fish in salt water during 2010 and 2011 increased. The production is expected to be restored to the 2006 level sometime between 2013 and 2015.

Atlantic Salmon production in Chile, 2000-2012

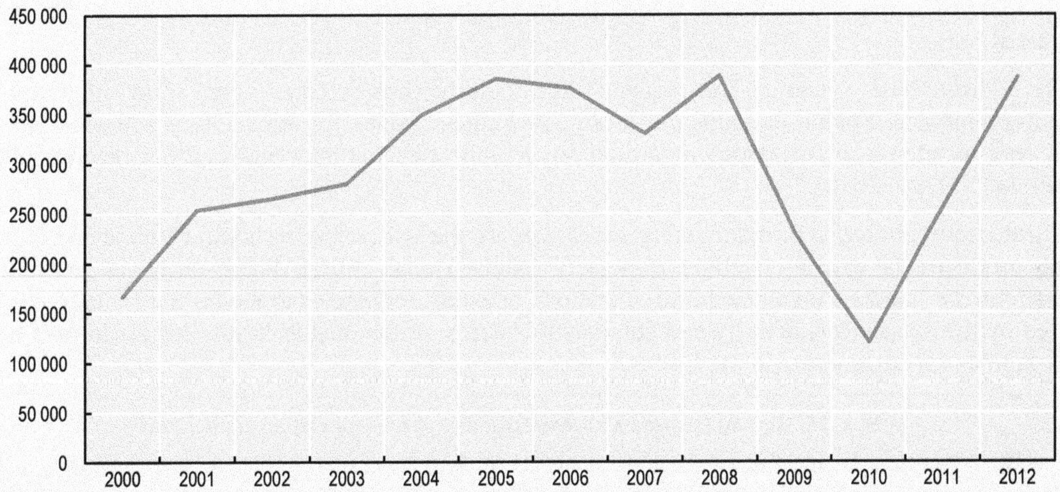

Several lessons have been learned from this crisis including: (1) development of R&D programs to provide timely information to support effective regulations and enforcement; (2) development of a biosecurity system covering the entire value chain; (3) understanding of the dynamics and biological carrying capacities; (4) establishment of effective zone management programmes; (5) reduction in drug treatments; and (6) maintaining good communication between industry stakeholders and government.

Source: Chilean Government, Under-secretariat for Fisheries and Aquaculture.

Box 3.5. The revival of tiger shrimp industry from white spot disease: The Sri Lanka case

Since its inception in the 1980s, tiger shrimp (*Penaeus monodon*) farming has been the most lucrative commercial aquaculture activity in Sri Lanka, especially in the north western and eastern parts of the country. The industry recorded its peak performances in 2000 by producing 4 855 tonnes of farmed shrimps, earning USD 69.4 million in foreign exchange. There was little attention to biosecurity and sustainability.

When white spot disease occurred in 2004, the shrimp farming industry was severely affected with a production dropping by two-thirds (1 570 tonnes in 2005). The problem worsened because the majority of the farms depended on the Dutch Canal, a common source for water intake and discharge, which facilitated the spread of disease from farm to farm.

Although most Asian countries changed to *Penaeus vanna*mei farming to prevent the recurrence of white spot disease, the Sri Lankan government resisted moves by most farmers to change and introduced measures supported by a legal framework to revive the shrimp farming industry. These measures include zoning of the shrimp farming area into zones and sub-zones; introducing a crop calendar; formation of a farmer organisation for each sub-zone; introduction of best management practices; active involvement of farmer organisations for decision–making and for the implementation management measures; avoiding high stocking densities; monitoring shrimp hatcheries and quality of post-larvae; and screening of broodstocks and post larvae for white spot disease.

The adopted measures resulted in the revival of the shrimp farming industry. The volume of production has reached around 3 500~4 500 tonnes, and the percentage occurrence of white spot disease decreased from 9.2% in 2005 to 2.9% in 2010.

Space competition

Aquaculture requires space on land and in water to operate. As aquaculture expands, suitable sites are scarcer and this has become a constraint in many regions on further growth as other economic sectors, such as fisheries, recreation, transportation and energy production, compete for space with aquaculture. This takes place not only at sea but also in harbours and in inland water aquaculture where access to aquifers may be limited or where the carrying capacity of the land and water bodies has been exhausted.

To address these competing uses aquaculture should be considered as a part of an integrated spatial planning approach. Spatial planning can designate suitable zones for aquaculture and other sectors, a good way to address many issues regarding aquaculture development and conflict resolution among stakeholders (Díaz, 2010).

Furthermore, in terms of maximising social welfare the scarce space should be allocated to a sector which produces the greatest welfare to society (Nielsen et al. 2012). When stringent environmental regulations are imposed on aquaculture operations one way forward is to assess the social contribution created by the aquaculture activity and compare this across competing sectors. This could lead to better opportunities for aquaculture in the future.

Box 3.6. Moving towards a zoning structure in the Norwegian aquaculture

Production of farmed salmon in Norway has grown continuously over the course of a 40-year period. In 2012, production amounted to 1.25 million tonnes, a doubling only since 2006. With expanded production has followed an increase in the area allocated for salmon farming - from 9 km^2 in 2000 to 59 km^2 in 2011. Historically however, aquaculture sites were allocated by virtue of a case by case approach, meaning there was no master plan in place for the overall structure of aquaculture sites. A viable and efficient site structure is an essential element in mitigating environmental concerns related to salmon farming and to facilitate future growth. In addition, competition for space from different user groups such as recreational users, fishers and the petroleum industry has made it increasingly difficult for salmon farming companies to get access to new sites.

In order to ensure industry optimisation and sustainable growth the Norwegian government has sought to explore the possibilities of an efficient zoning structure for aquaculture. A zoning committee appointed in 2009 put forth its recommendations in 2011. Its main suggestion was to divide the coastline into production areas separated by corridors. Each production area should further be divided into at least four zones with coordinated smolt release and fallowing (rotating principle). This is believed to reduce disease outbreaks and help to better manage and implement current and future environmental indicators and sustainability goals. Several issues were raised during the committee's hearings. These included 1) knowledge gaps for establishing suitable production zones, 2) challenges for small farm owners located in only one or two zones, and 3) the municipalities' responsibility for spatial planning processes in coastal waters.

Source: Norwegian Ministry of Fisheries and Coastal Affairs (2012).

With growing pressure on land and marine resources, a sound approach to spatial allocation is vital to avoid conflict and wasted opportunities. The report *Integrated Ocean Management and the Fisheries Sector: Interactions Economic Tools and Governance Structures,* discussed at the 107th Session of COFI, considered how to address the problems associated with user conflicts and identify the best ways to deal with them (Charles, 2011) (Box 3.6).

Externalities from other sectors

Aquaculture may suffer from externalities induced by other sectors. Since water is of utmost importance for aquaculture activities, other activities that deteriorate water quality or decrease water availability may produce negative externalities on aquaculture operations. In fact, there are increasing impacts from land based activities, such as agricultural run offs, municipal sewage and industrial waste, which deteriorate water quality and that can have potentially negative impacts on aquaculture, both in inland and marine-based farming. Agricultural run-offs are generally the greatest contributor to eutrophication in many countries (Díaz, 2010).

There are 415 eutrophic and hypoxic coastal systems in the world of which 169 areas are hypoxic areas. Particularly sensitive zones include the Gulf of Mexico and east coast of the United States, north-east Atlantic and seas around United Kingdom and the southern coast of Japan and Korea (Selman et al., 2010; see www.wri.org/map/world-hypoxic-and-eutrophic-coastal-areas).

Box 3.7. Dealing with externalities from other sectors: the French experience

In France, the main externalities on aquaculture operations from other sectors are due to the competition for water use (agriculture, livestock, industry, urbanisation, tourism) and the competition for the use of sites (urbanization and marine tourism). The problem of pesticides and other chemicals is no longer a priority with the introduction of strong European and national regulations.

The Arcachon basin is located in the west-southern coast of France and a multipurpose area used for: oyster farming with 15 000 tonnes of annual commercial production and the main region for spat production in France; tourism and urbanisation with 6 million nights during the summer season (equivalent to 250 000 inhabitants: three times the normal population of this area); sailing boats with 12 000 boats moored inside the basin; and a paper mill, which is one of the most important in France, uses the neighbouring pine tree production (1 million hectares of pine trees).

The main externalities on aquaculture operations identified were: a competition for the use of the surface area of the Basin (15 000 ha at high tide, but 5 000 ha at low tide); the risks due to urban and industrial pollution; and the access to the sea (physical access and cost of coastal land).

To preserve the aquaculture activities against these externalities, a regional marine spatial planning called the *Schéma de Mise en Valeur de la Mer* (SMVM) was implemented. The SMVM of the Arcachon Basin determines the role of the different sectors in the maritime and coastal area in order to define the compatibility between different uses and specific protective measures. It was developed under the responsibility of the state, in close consultation with the various stakeholders and sets guidelines for the development, protection and equipment that will shape the future of the Basin. It has the same legal status as a territorial development law. Another important action taken was construction of a great circular collector to receive all wastewater from urban origin and also from the paper mill. All these effluents were sent to the open sea after treatment.

Unlike the success story in the Arcachon case, externalities imposed on trout farms in Brittany failed to be appropriately addressed.

Brittany was the most important French region for the production of trout during the 1970s, producing about 15 000 tonnes. But the increasing production of pigs in Brittany (14 million pigs annual production) has had an important impact on trout production. In addition, the development of pig production has an impact on the water quality of the coastal zone. The main problem is to take into account the 24 000 direct jobs induced by the pig production.

Presently, no solution has been found and trout production has decreased by 40% during the last 20 years. This could have been avoided with developing planning tool which could determine the role of different sectors of the maritime and coastal area in defining the compatibility between different uses and specific protective measures.

Since multiple externalities and economic activities are involved it is not possible to correct these by addressing aquaculture alone. Rather, co-ordinated regulation of externalities among different sectors can internalise different players' externalities in their management decision and ensure the highest possible value is produced (Nielsen et al., 2012). For example, transferable discharge permits could be applied to emissions from both the aquaculture and agriculture sectors. A focus on coherence and a willingness to co-operate across government agencies/ministries and with a wide variety of stakeholders who have an economic interest in this shared resource is necessary for this process to succeed.

Addressing multiple externalities can increase the complexity of legal and institutional arrangements. The one-stop shop approach taken in Norway and the state of Michigan in the United States can reduce the burden of administrative process by providing concerted and streamlined services through one channel (OECD, 2010a).

In France, application of spatial planning frameworks to address externalities has produced mixed results (Box 3.7). This experience shows that spatial planning and bringing together different stakeholders can be more difficult that it seems, especially when the potential value stakes are high.

Using resources efficiently

The use of wild feed in aquaculture

Aquaculture is the biggest fishmeal and fish oil consumer and it is estimated to consume more than 50% and 80% respectively of the world fishmeal and fish oil production (Hasan and Halwart, 2011). In 2006, about 37% of global aquaculture production (19.3 Mt) relied on small pelagic fisheries for its feed (Tacon and Metian, 2009). Continued growth will put additional pressure on forage fisheries from which fishmeal and oil is produced. The potential existence of a fishmeal trap, where aquaculture growth is limited because of the lack of fish used for fishmeal and oil production is a major concern, especially when carnivorous species such as salmon are concerned. Sustainable management and conservation of forage fish stocks is central and vigilance against overharvesting is required. Concurrently alternative feed ingredients need to be developed. Innovation policies are of key importance to achieve this.

For some, using wild fish to produce farmed fish raises food security and ethical issues. In addition, as other resources such as grains have been increasingly used as substitutes for fish meal and fish oil, similar question arise for other ingredients of fish feeds, i.e. alternative uses of soy, colza, etc. The objective is to find additional sources of feed (whether terrestrial or marine) that are managed sustainably.

However, there is room for optimism. The total amount of wild capture fish used for reduction (transformation) to fish meal and oil has remained stable over the last three decades while aquaculture production has substantially increased over the same period (Tacon and Metian, 2009) (Figure 3.3). The use of fish meal in compound aquaculture feeds has been lower than predicted between 1997 and 2007 (Welch et al., 2010) (Figure 3.4).

As the aquaculture industry grows, new substitutes for fish meal and oil will be needed, and for more species. This means a need for more research and innovation in ingredient substitution and improved feeding systems. (Bostock et al., 2010). The "fish-in fish-out ratios" for nearly all species fell between 1995 and 2006 (Table 3.6). While biomass transfer efficiency between trophic levels of fish is generally around 10% in natural environments, all farmed species cited in the table have greater transfer efficiency than their counterparts in the wild. In case of salmon, improvements after 2006 are particularly notable (Figure 3.5).

In the near term, supply of fish-based feedstocks will probably not be the limiting factor for aquaculture growth. The use of fish meal and fish oil per unit of output has decreased, mainly because of industry's keen interest and investment to find substitutes in response to rising prices and social pressure for improved sustainability (Bostock et al. 2010). The industry is expected to continue to find ways to drive down fishmeal and oil levels in the future (Welch et al., 2010). In particular, there have been major achievements in reducing the fishmeal and oil component in the salmon industry

(Figure 3.6). In addition, about 25% of fish meal and oil sources are now provided by processing waste, i.e trimmings and waste water from fish processing (Jackson, 2010). As it is now technically feasible to produce carnivores such as salmon using low levels of fishmeal and oil, the quantity of Omega 3 oils available in the fish is becoming an issue. Ultimately, two market segments for carnivorous species may develop: a premium one with high levels of Omega 3 and another low price, but poor in Omega amino acids.

Table 3.6. Calculation of pelagic forage fish equivalent per unit of cultured species groups

	1995	2005	2006
Salmon	7.5	5.4	4.9
Trout	6	4.2	3.4
Eel	5.2	4	3.5
Marine fish	3	2.1	2.2
Shrimp	1.9	1.7	1.4
Freshwater crustaceans	1	0.9	0.6
Tilapia	0.9	0.6	0.4
Catfish	0.4	0.6	0.5
Milkfish	0.4	0.2	0.2
Non-filter feeding carp	0.2	0.3	0.2
Total major fed species	1	0.9	0.7

Source: Tacon, A.G.J. and M. Metian (2008), "Global Overview on the Use of Fish meal and Fish oil in Industrially Compounded Aquafeeds: Trends and Future Prospects", *Aquaculture*, Vol. 285.

Figure 3.3. World fish meal and fish oil production from 1976 to 2006

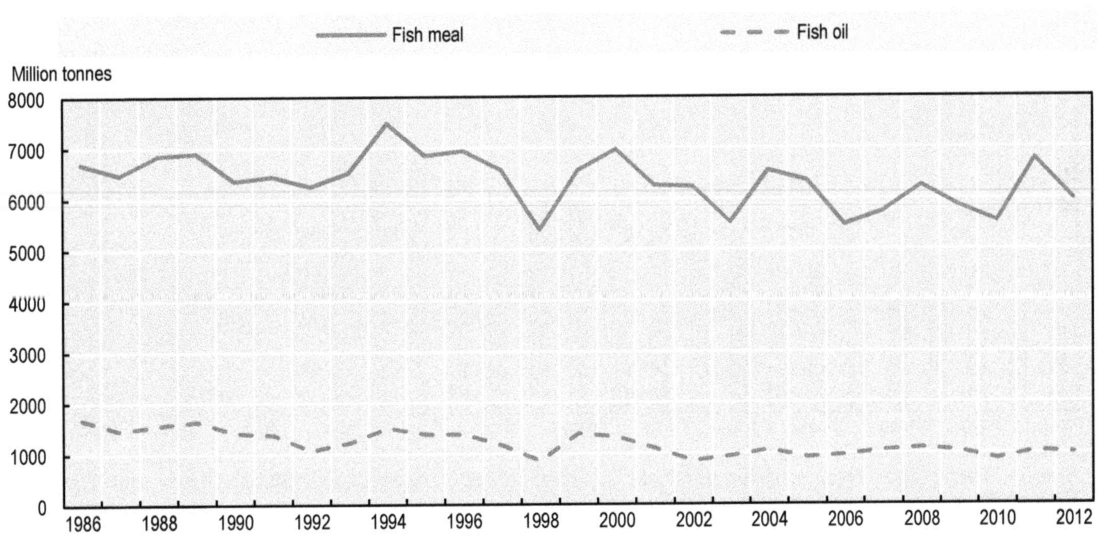

Source: OECD/FAO (2013), *OECD-FAO Agricultural Outlook 2013*, OECD Publishing, Paris.
DOI: http://dx.doi.org/10.1787/agr_outlook-2013-en.

Figure 3.4. Fishmeal use: Past projections and current trends

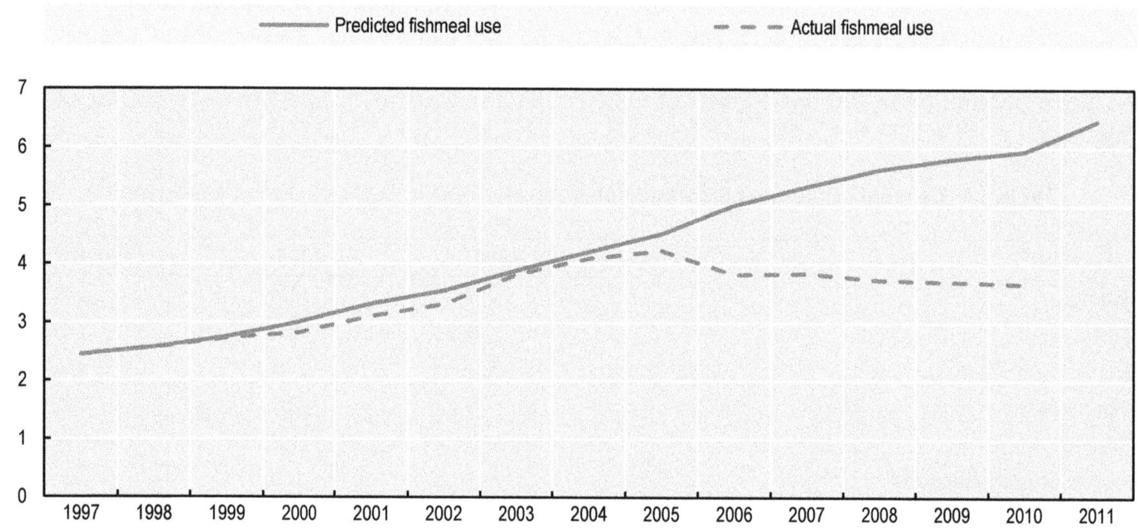

Source: Author's calculations updating Kristofersson , D. and Anderson , J. L. (2006), "Is there a relationship between fisheries and farming? Inter-dependence of fisheries, animal production and aquaculture." *Marine Policy*, vol. 30:721–725.

Figure 3.5. Estimated global use of fish meal and oil by the salmon farming industry projected to 2020

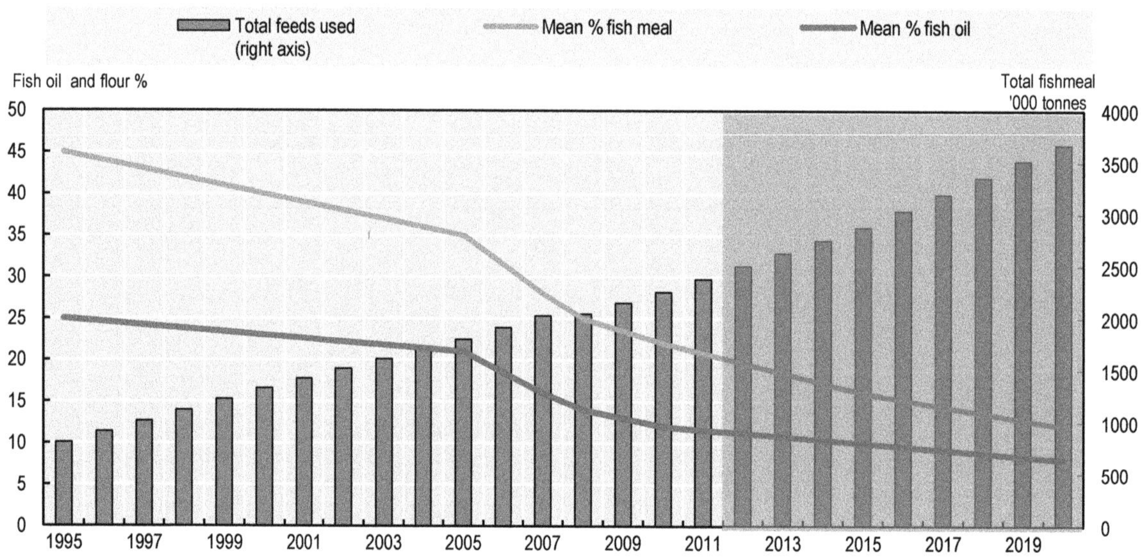

Source: Tacon and Metian (recited Bostock, J., et al. (2010), "Aquaculture: Global Status and Trends", *Philosophical transactions of the Royal Society,* Vol. 365).

Energy intensity of aquaculture production

Energy use in aquaculture depends on many factors including farming methods (i.e. cage *vs*. recirculation, intensive vs. extensive farming), species cultured (i.e. carnivorous *vs*. herbivorous vs. non-feed species such as seaweed and mollusc), regions and energy mix etc. In general, recirculated aquaculture systems, intensive aquaculture systems and carnivorous species farming are likely to demand more energy than cage systems, extensive systems, herbivorous and non-fed farming systems (e.g. oysters, mussels) respectively.

Ayer and Tyedmers (2008) carried out life cycle assessments of four different salmon farming systems in Canada and found that there were substantial differences in energy use among four production modes (Table 3.7). Feed dominates cumulate energy demand (CED) both in conventional marine net-pen system (87%) and marine floating bag system while electricity and fuel dominates CED in land-based saltwater flow-throw system and land-based freshwater recirculating system (90%), even though they use an equivalent amount of feed.

Marine net-pen outperforms closed-containment systems in energy efficiency mainly because marine net-pens utilise ecosystem services (i.e. ocean currents and tidal action) in supplying fresh sea water and dissolved oxygen and flushing out wastes while closed-containment systems have to provide those using energy. Gronroos et al., (2006) studied Finnish rainbow trout farms and found that there were also big differences in energy use among different production modes. One interesting observation in their study is that while feed production dominated total energy use in a net-pen system, it accounts only a fifth of total energy use in a land-based recirculating system. The high cost of energy in many areas serves as a significant impediment for the development of closed recirculating systems. A recent study (Feasibility Study of Closed-Containment Options for the British Columbia Aquaculture Industry, 2010, DFO, Canada) concluded that closed recirculating systems are not suitable for salmon farming as the high operating cost (mostly energy) made it economically unviable. This may change in the future but for now such systems seem viable for high-value species only or in areas where energy cost is very low. However, closed-containment systems have other environmental advantages such as less eutrophication and escapees and so properly valuing these environmental costs is important to choosing the ideal system.

Energy use among the same production systems with the same species can vary across regions. Pelletier et al., (2009) did life cycle assessment on salmon farming systems (marine net-pen) in Norway, the United Kingdom, Canada and Chile and found that there were also substantial differences in material/energy use and environmental impacts across countries (Table 3.8). Feed production had a dominant influence on all categories analysed, for example, feed accounts for 93% of farm-gate cumulative energy use and 94% of global warming and acidifying emissions. Despite the fact that they are producing essentially the same product, there were substantial differences of between 40%~80% in on farm level energy use compared with Norway, the most efficient one. The big difference in farm-gate cumulative energy use can be attributed to feed milling (21~54% difference compared to Norway) and food conversion ratio (19~35% difference). Such major differences in the same production system across countries suggest there is significant room for improvement especially through efficient feed sourcing and better feed conversion ratios. It also highlights the considerable cost efficiency of some species across countries and hence competitive advantage.

Table 3.7. Impact of production from four different aquaculture systems

Per tonne live-weight fish

	ABD (kg Sb eq)	GWP (kg CO2 eq)	HTP (kg 1,4-DIB eq)	MTP (kg 1,4-DIB eq)	ACD (kg SO2 eq)	EUT (kg PO4 eq)	CED (MJ)
Net-pen	12.1	2 073	639	822 000	17.9	35.3	26 900
Bag	13.9	2 250	840	574 000	18.0	31.9	37 300
Land-based flow-through	38.1	5 410	2 570	3 840 000	33.3	31.0	132 000
Land-based recirculating	72.5	10 300	54 380	6 150 000	63.4	11.6	233 000

Notes: Systems analysed with all closed-containment systems assumed to be operating on the average Canadian electricity mix.

ABD=abiotic depletion; GWP=global warming potential; HTP=human toxicity potential; ACD=acidification; EUT=eutrophication; and CED=cumulative energy demand.

Source: Ayer, N.W. and P.H. Tyedmers (2008), "Assessing alternative aquaculture technologies: Life cycle assessment of salmonid culture systems in Canada", *Journal of Cleaner Production*, doi:10.1016/j.jclepro.2008.08.002.

Table 3.8. Input use for salmon farming and salmon feed milling in Norway, the United Kingdom, Canada and Chile, 2007

	Norway	UK	Canada	Chile
		Inputs per tonne of salmon		
Feed (t)	1.103	1.331	1.313	1.493
Feed transport (t-km)	290.3	321.7	316	298.7
Smolts (kg)	17.4	22.2	16	15
Smolt transport (t-km)	1.2	3.9	3.2	3
Total on-farm energy use (MJ)	646.8	904	933.7	1199
Farm-level emissions (kg N/P)	41.1/5.2	58.7/8.5	51.4/13.6	71.3/12.6
		Inputs per tonne of feed		
Energy for feed milling (MJ)	902.6	1090.1	1393.2	1118.7
		Feed composition (%)		
Crop-derived meals/oils	35.3/6.1	32.3/1.1	43.4/5.1	36.9/5.8
Animal-derived meals/oils	-	-	16.8/3.1	15.1/0
Fish-derived meals/oils	33.1/25.5	40.5/26.1	20.9/10.7	25.1/17.1

Source: Pelletier, N. et al. (2009), "Not all salmon are created equal: life cycle assessment (LCA) of global salmon farming systems", *Environmental Science & Technology*, Vol. 43.

In conclusion, there is substantial scope for reducing energy use in aquaculture in that there are significant differences in energy use as well as other environmental aspects across countries, different production modes and different species. Innovations in feed ingredients and farming facilities, introducing best management practices in feeding and introducing more efficient and efficient regulations can help realise this potential. Combining aquaculture and new energy sources (such as windmill or sun energy) may also provide green growth opportunity in terms of energy use in the near future.

Enable change

Governance

Competent governments' involvement and support for aquaculture through various policies was identified as a key factor in explaining the reason for differences in aquaculture development among seven Southeast Asian aquaculture countries (Cambodia, Indonesia, Malaysia, Myanmar, the Philippines, Thailand and Viet Nam). The capacity of public officials in a country mattered for the success of aquaculture development (Hishamunda et al., 2009). A recent study (Fukuyama, 2013) argues that the quality of government is the result of an interaction between capacity and autonomy, which is an inversed U curve with bureaucratic autonomy on the horizontal axis and the curve shifts upward right as capacity of bureaucrats increase. As the capacity of bureaucracies increase, they can be granted greater autonomy which improves responsiveness, flexibility and innovation in problem solving. Thus investments in human and institutional can help spur green growth, including in aquaculture.

In addition to a long tradition of seaweed aquaculture for food consumption, Korea has proactively invested in innovative research projects together with academia to create new opportunities for seaweed aquaculture. Since 2006, several national seaweed projects have been proposed such as using seaweed to reduce greenhouse gas emissions (GHG), to develop biomass energy sources and to develop environmentally friendly fish cage (Box 3.8). This case shows how government's vision and action can support green growth in aquaculture through R&D investment.

Although growth in EU aquaculture production has been small over the last decade, aquaculture has the potential to boost growth and jobs both in coastal and inland areas of the European Union. The European Commission recently issued strategic guidelines to promote aquaculture development in the European Union, identifying key challenges facing the sector and policy measures to address them (Box 3.9).

Box 3.8. Seaweed aquaculture for green growth in Korea

A five-year project on the reduction of greenhouse gas (GHG) emissions using seaweed was conducted starting in 2006. The project was funded by the Ministry of Maritime Affairs and Fishery and later the Ministry of Land, Transport and Maritime Affairs of Korea. The project uses innovative research on seaweed to develop new baseline and monitoring methods for the CDM and Project Design Document (PDD) of the Kyoto Protocol. Concurrently, members of the project and the APPA have been playing a key role in obtaining international recognition of seaweed as a GHG sink (Kang *et al.*, 2008; Chung *et al.*, 2011). The new concept of the Coastal CO2 Removal Belt (CCRB) has been established for natural and/or man-made plant communities in the coastal region to accomplish CO2 removal in the same way as a forest and that can be implemented on various spatial-temporal scales. About 10 tonnes of CO2 per ha per year could be removed in the pilot scale CCRB farm with perennial brown alga *Ecklonia*, estimated by the biomass increment and decrease in the dissolved inorganic carbon in the water column.

The biomass project was conducted by five Ministries of the government with a funding of USD 178 million. Each ministry has different projects but the final goal is to develop biomass energy sources, find bio-materials and reduce CO2. The Ministry of Food, Agriculture, Forest and Fisheries (MIFAFF) undertakes the "*Seaweed biomass production for green energy project*" with an objective of finding a substitute for 30% of gasoline consumption and increasing employment in this sector by 40 000 jobs by 2020.

An Integrated multi-trophic aquaculture (IMTA) project was modified as an environmentally friendly fish cage to include a tourist view point located on the eastern coast of Korea. The fish cage pilot was built in 2011 and the project will run from 2011 to 2013.

Box 3.9. Strategic guidelines for the sustainable development of EU aquaculture

The European Commission in close consultation with stakeholders identified four main challenges facing the sector and presented proposals to address those challenges to unlock the potential of EU aquaculture.

- Simplify administrative procedures to reduce red tape and uncertainties. For example, licensing procedure in several EU member states often take 2-3 years while average licensing time for aquaculture farms in Norway has been reduced to 6 months from 12 months since the introduction of a single contact point. The burden of administrative costs (time) is substantial to aquaculture farmers given that most aquaculture farmers are SMEs.
- Introduce co-ordinated spatial planning to secure sustainable development and growth of aquaculture. Spatial plans can help reducing uncertainty, facilitating investment, identifying suitable sites and taking into account environmental aspects.
- Enhance the competitiveness of EU aquaculture with an improved market organisation and structuring producer organisations.
- Promote a level playing field for EU operators by exploiting their competitive advantages of high quality products complying with high environmental, animal health and consumer protection standards. New labelling and voluntary certification scheme can serve for this purpose.

The guidelines also presents a new governance to support EU aquaculture:

- Multiannual national strategic plan based on EU strategic guidelines for the promotion of sustainable aquaculture.
- Complementarity with European Maritime and Fisheries Fund.
- Exchange of best practices through peer review seminars to develop a mutual learning process.
- Aquaculture advisory council to utilise the knowledge and experience of all stakeholders for evidence-based decisions.

Source: European Commission (COM(2013)229/final), http://ec.europa.eu/fisheries/cfp/aquaculture/official_documents/com_2013_229_en.pdf.

Stakeholder participation

Broad and active stakeholder participation in policy-making, planning and management is expected to produce more effective and informed policy responses while facilitating implementation. In addition, stakeholder participation can give policies more legitimacy and augment trust in government. Stakeholder participation makes it easier to develop more effective policies by bringing more information and experience into the process, building support while reducing opposition and conflicts among stakeholders (Sen, 2001). Therefore stakeholder participation can help make policies more effective, improve performance in production and with regard to environmental impacts as well as help on the public perception of the aquaculture industry.

Many countries encourage stakeholder participation in their aquaculture management process though the level of participation may vary. In Sri Lanka shrimp farmers are actively involved in the formulation and implementation of the crop calendar through their aquaculture organisations to avoid high risk periods in respect of stress factors, sustainability of water quality and post larvae quality. Not only has this process resulted in a more realistic calendar by taking into account more site-relevant information and experience but also helped implementation by increasing farmers' understanding and acceptance of the process.

The Japanese aquaculture industry put a priority on increasing production until the late 1990s. This resulted in excessive culturing density and excessive feeding potentially causing environmental and economic damage. In 1999, the Sustainable Aquaculture Production Assurance Act was enacted which emphasises the importance of the role of fisheries organisations and other local organisations. By the law Fisheries Cooperative Associations are asked to prepare a "Plan concerning the improvement of aquaculture ground" including measures for the prevention of infectious disease of aquatic animal and plants. While the Plan is authorised by the competent Prefectural Governor, Fisheries Cooperative Associations play an important role in the aquaculture management process.

Gender equity

Women have an important role in fisheries and aquaculture. Worldwide, fish products support the livelihoods of 520 million people many of whom are women (FAO 2009, WorldFish Center, 2009). Women are the main providers of health care and contribute to food security and poverty alleviation at the household level. Furthermore their roles at a macro level are significant contributing 47% of the labour force in the fisheries sector based on available statistics and case studies (FAO, World Bank and WorldFish, 2009). If statistics for aquaculture were included, these records could be higher.

Box 3.10. Gender inequality issues in aquaculture

In aquaculture, women are engaged in all activities along the value chain while in capture fisheries their most prominent role is documented in post-harvest, processing and marketing. Aquaculture, especially in Asia shows that women's labour contribution is often greater than men's. Women are reported to constitute from 33% of the rural aquaculture workforce in China to between 42 and 80% in freshwater and cage culture in Indonesia and Viet Nam. Hence, women could gain considerable economic independence and empowerment through their work in aquaculture. In this context, microcredit programs in Asia for example have promoted women's access to economic opportunities and employment.

The main issues of gender inequality in in the fisheries and aquaculture sectors include:

- Unrecognised, undervalued and/or unpaid contribution when working for a family enterprise
- Exclusion from decision-making bodies
- Limited access to credit, training, storage facilities and new technologies
- Lack of Information and data on gender

The following actions are required to in order to enhance gender equality in aquaculture:

- Providing working women with a legal/professional status and rights
- Support for improvement of women's work (training, funding opportunities)
- Development of research for gender and policy
- Recognising contribution (direct and indirect) of women to the economy
- Improve inclusion of women in decision making processes

Source: FAO (2011), "The role of women in agriculture", prepared by SOFA Team 2 and C. Doss in World Fish Center (2009), *Gleaner, fisher, trader, processor: Understanding gendered employment in the fisheries and aquaculture sector*, Food and Agriculture Organisation Publications, Rome.

Although women are more involved in aquaculture activities than in capture fisheries, gender inequalities remain important in both sectors. The inequalities may be more significant in aquaculture given the strong involvement of the women in the sector (Box 3.10).

Innovation

An important part of any green growth strategy is promoting innovation. Ideally, innovation can lead to more production using the same level of natural resources while reducing negative impacts on the environment (OECD, 2011b). For this reason, most governments provide incentives for innovation by rewarding firms for undertaking R&D. Government expenditures on research and development is also common as innovations can bring broad benefits and can be considered a public good. Small scale farmers form the majority of producers in most countries, but often lack resources to innovate on their own. An industry-wide approach to innovation activities in aquaculture (through industry organisations, fish farmer organisations, specialised university laboratories, etc.) and a strategy for the diffusion of the results of the innovation can bring broad benefits. Also, public intervention may be needed to "kick start" innovations.

Technological innovations have played a very important role for growth in every aspect of aquaculture operations, such as control of life cycle, feed, facilities, reducing negative environmental impacts, to name a few. Asche (2008) summarised how innovations have contributed to aquaculture development: control over biological processes allowed systematic research which provided productivity improvements and potential for specialisation, which expedited many innovations in aquaculture. For example, the single innovation of a vaccine in 1991 reduced production cost by 5-10% and contributed to a dramatic decrease in antibiotic use in Norway while production volume increased more than 15 times (from 47 200 to 744 222 tonnes) between 1987 and 2007 (Figure 3.6).

Figure 3.6. Use of antibiotics in the Norwegian aquaculture industry

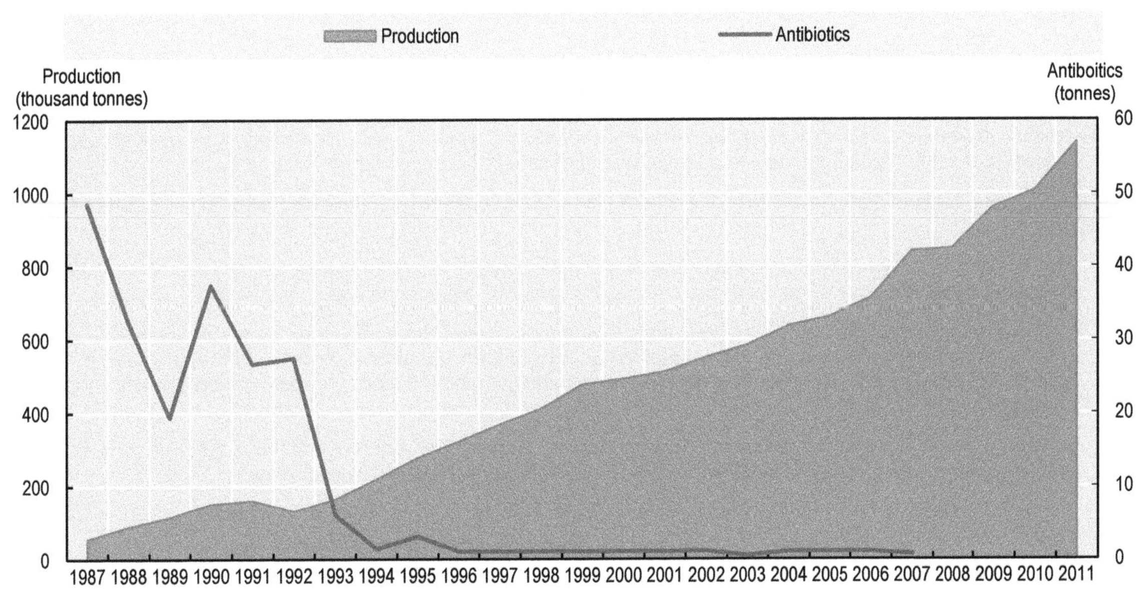

Source: The Norwegian Directorate of Fisheries and the National Health Institute.

As of 2010, a number of Danish trout farms adopted new recirculation systems. The most advanced model fish farm re-circulates at least 95% of water, reduces water intake about 15-25 times, and reduces discharge of total nitrogen, total phosphorous, and organic material by respectively 36%, 62% and 94% compared to the traditional farms (Jarlbæk and Børresen, 2011). In Denmark the most important factor limiting growth in trout aquaculture is nitrogen and phosphorous discharges in inland waters. Hence, in order to limit this pollution water purification processes/systems have been developed in freshwater aquaculture. The main reasons for changing to this innovative system are strict environmental regulations combined with strict regulations of using weirs subject to maximum feed quotas, statistical standard for nitrogen and phosphorous organic matter, minimum level of oxygen in the outlet water and a limit on water intake (Jarlbæk and Børresen, 2011). Thus, on the one hand, the strict regulations have hindered the aquaculture development in Denmark, but it has also accelerated innovation in the sector. Several initiatives have been launched with the aim to increase the Danish aquaculture production in general such as exploring new concepts for aquaculture production with reduced environmental footprint.

The Norwegian salmon industry has sought to reduce production costs in many ways. Optimisation of holding facilities and handling and feeding equipment have contributed their share of the success in reducing production costs. However, the Norwegian selective breeding program1 for salmon starting in the 1970s and a lowering of the fish feed conversion ratio combined with a lower use of fishmeal and fish oil in the feed have been the most important contributing factors to this success (OECD, 2010). The supply industries and the government have played important roles in this process. In Norway, salmon farmers historically had little resources for R&D and so were dependent on their suppliers for this type of research (Box 3.11).

Box 3.11. Innovations strategies and green growth: The case of Norwegian salmon farming

Innovation in all areas related to salmon farming has been the main reason for the growth in Norwegian production of salmon since the late 1960s. A paper by Frank Asche, Kristin Roll and Ragnar Tveterås argue that there is a direct relationship between R&D, innovation and productivity growth in Norwegian salmon farming where successful R&D results in innovation has led to major productivity growth (Asche et al., 2012). The supply industry and the government have played a vital role in this process. Three historically important sources of productivity growth have been identified: 1) innovations in key technological areas; 2) increased know-how in all areas; and 3) economies of scale throughout the value chain (Asche et al., 2012).

Salmon farming firms can be listed under one of four categories, depending on their innovation strategies (Aslesen, 2007):

- The family firm is a small family-owned and family run company with little resources for R&D. Companies of this category do not have a real innovation strategy and rely on experience-based knowledge.
- The coastal enterprise is a more professionally-run company than the family firm but has no interest in doing R&D. Companies of this type are mainly concerned with efficiency and cost control and they pursue an "anti-innovation" strategy by consciously avoiding new technologies until they have been proven to work by other companies.
- Research-based entrepreneurs control parts of the value chain that require continuous R&D, pursue radical innovations and are happy to share their innovations with other companies in the cluster.
- A company which is part of the science-based process industry is a fully integrated company which is able to apply its skill and capabilities to build a competitive advantage based on innovation.

Historically, most salmon farming companies in Norway have been small family-run firms that are dependent on their suppliers for innovations and new technologies. As of 2007, only a few companies could be categorised as research-based entrepreneurs or part of the science-based process industry. A number of companies were still pursuing anti-innovation strategies.

As most salmon farming companies lack the means and capabilities to appropriate and internalize the benefits of their R&D-activities, there are disincentives for salmon farming companies to take on large R&D investments. As such government funded research, which historically has been integral to the innovation system of Norwegian salmon farming, will continue to play an important role in the future. However, Asche et al. (2012) argue that productivity growth in salmon farming has stalled since the mid-1990s, coinciding with a drop in R&D intensity. Asche et al. (2012) hold that salmon farming companies themselves may need to increase their R&D-capabilities if the industry is to produce the kind of incremental and especially radical innovations that has been driving productivity growth in the past.

Box 3.11. Innovations strategies and green growth: The case of Norwegian salmon farming (*cont.*)

Production cost per kilo and production of salmon, tonnes

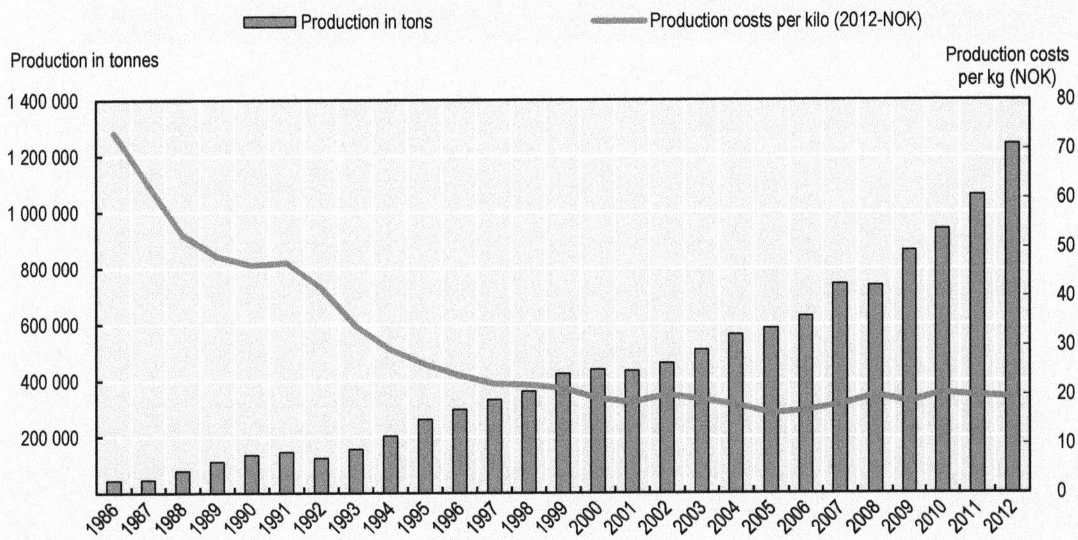

Source: Norwegian Ministry of Fisheries and Coastal Affairs (2012).

Deliver on green growth

Many countries have already incorporated some green growth measures in their regulatory frameworks. However, little is known about the effects of adopting green growth measures in aquaculture policies and whether they have fostered or hindered the competitiveness of the industry in the global market for fish and fish products. Since about 50% of fish and fish products in the world are traded, the impact on competitiveness can have potentially important implications. While there is not much information available for in-depth analysis, there are several cases that provide some indication of effects of green growth policies on competitiveness in aquaculture development.

Denmark introduced a farm-specific feed quota system in the 1990s to prevent eutrophication and pollution from aquaculture production. Since then, Danish aquaculture production has decreased from 44 730 tonnes (USD 145 million) in 1995 to 39 507 tonnes (USD 136 million) in 2010 (Figure 3.7). The regulation has been criticised because of its inefficiency and lack of flexibility, which has led to the sub-optimal regulation of the sector (Nielsen, 2012). There was a rapid growth in production until 1990; since then the production has stagnated and later decreased. Nielsen (2012) showed that *changing this regulation to individual transferable quotas on nitrogen could increase Danish aquaculture production by 16% to 55% and profitability by five to ten times while keeping the current pollution level.* A new regulation for fish farming entered into force on 15 February 2012 giving the option to fish farmers to voluntarily choose regulation based on discharge with among others specific nitrogen limits instead of feed quotas. It is likely that this shift in regulatory framework will make the Danish trout industry more competitive. Other recommendations with the aim to encourage sustainable growth are still under consideration.

Chilean responses to the ISA crisis provide some lessons on the results of green growth policies. The ISA crisis led to a substantial decrease in the production of Atlantic salmon, a significant fall in the number of Atlantic salmon farms and a significant loss in direct and indirect jobs (Chile, 2012) (Figure 3.9). Though not fully recovered yet, there are signs of recovery and effects of the green growth policy. Production began to increase in 2011 followed by a reduction in the use of antibiotics which in 2008 was 350 times more than Norway per kilo of salmon produced (Chile, 2012; Asche et al., 2010).

Figure 3.7. Danish aquaculture production between 1980 and 2010

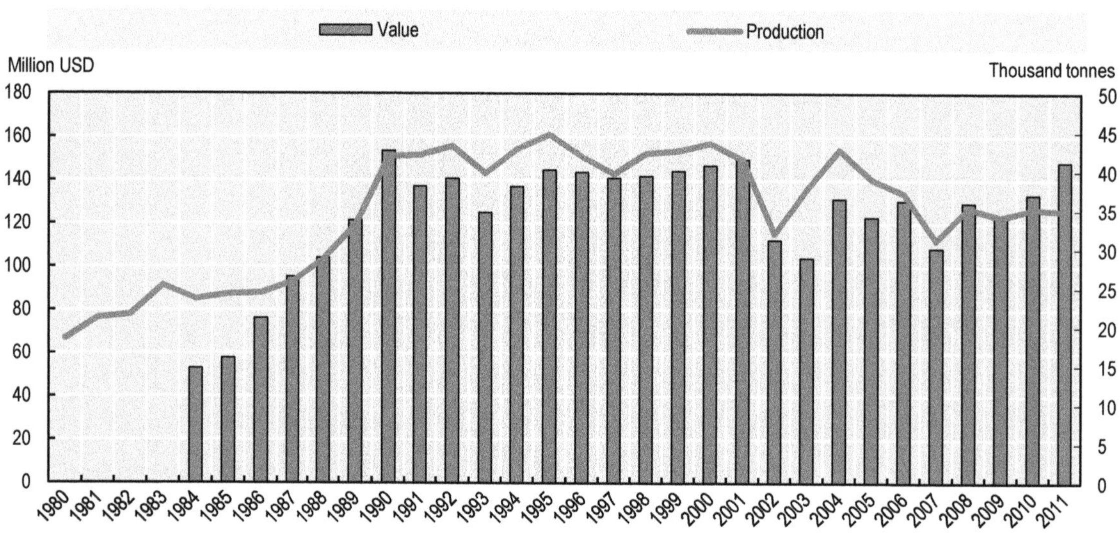

Source: FAO (2014), *Fishery and Aquaculture Statistics. Global production by production source 1950-2012* (FishstatJ). In: FAO Fisheries and Aquaculture Department [online or CD-ROM]. Rome. Updated 2014. http://www.fao.org/fishery/statistics/software/fishstatj/en.

Figure 3.8. Atlantic salmon production among major countries

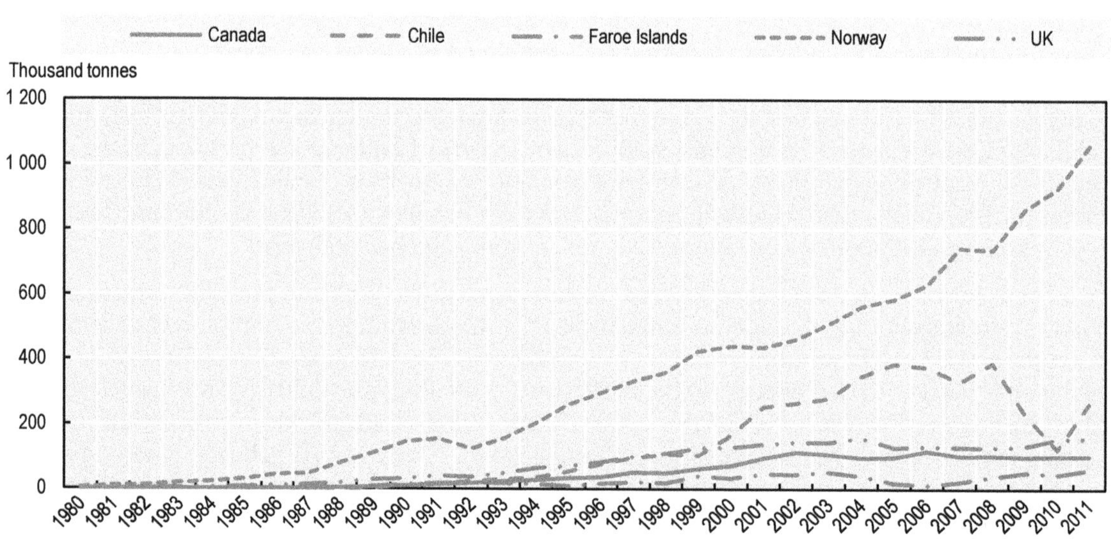

Source: FAO (2014), *Fishery and Aquaculture Statistics. Global production by production source 1950-2012* (FishstatJ). In: FAO Fisheries and Aquaculture Department [online or CD-ROM]. Rome. Updated 2014. http://www.fao.org/fishery/statistics/software/fishstatj/en.

Box 3.12. Strategic guidelines for the sustainable development of EU aquaculture

The European Commission in close consultation with stakeholders identified four main challenges facing the sector and presented proposals to address those challenges to unlock the potential of EU aquaculture.

- Simplify administrative procedures to reduce red tape and uncertainties. For example, licensing procedure in several EU member states often take two-three years while average licensing time for aquaculture farms in Norway has been reduced to six months from 12 months since the introduction of a single contact point. The burden of administrative costs (time) is substantial to aquaculture farmers given that most aquaculture farmers are SMEs.
- Introduce co-ordinated spatial planning to secure sustainable development and growth of aquaculture. Spatial plans can help reducing uncertainty, facilitating investment, identifying suitable sites and taking into account environmental aspects.
- Enhance the competitiveness of EU aquaculture with an improved market organisation and structuring producer organisations.
- Promote a level playing field for EU operators by exploiting their competitive advantages of high quality products complying with high environmental, animal health and consumer protection standards. New labelling and voluntary certification scheme can serve for this purpose.

The guidelines also presents a new governance to support EU aquaculture:

- Multiannual national strategic plan based on EU strategic guidelines for the promotion of sustainable aquaculture.
- Complementarity with European Maritime and Fisheries Fund.
- Exchange of best practices through peer review seminars to develop a mutual learning process.
- Aquaculture advisory council to utilise the knowledge and experience of all stakeholders for evidence-based decisions.

Source: European Commission (COM(2013)229/final), http://ec.europa.eu/fisheries/cfp/aquaculture/official_documents/com_2013_229_en.pdf .

A compelling case has been put forward by Pelletier et al. (2009) which have analysed carious factors through a life cycle assessment process of salmon farming across the four major producers Norway, United Kingdom, Canada and Chile. Their analysis suggests that policies that protect the environment do not lead to cost disadvantages for producers in those countries, with Norway being a good example of this.

As highlighted above and as shown in the case studies presented at the December 2012 *Korean Yeosu Workshop on Aquaculture and Green Growth*, aquaculture has a strong potential to deliver green growth. Many countries are now looking at regulatory reforms to help deliver on this promise. New technologies can reduce the environmental impact of production. Market based instruments have been successfully introduced in some jurisdictions (e.g. replacing allowable feed quotas with tradable N emission permits). New innovations will continue to emerge, and the cost of alternative production methods will continue to decline as technologies mature. A change in focus from regulating inputs like feed quantities or production systems to outputs like emissions and effluents will promote innovation. A move towards aquaculture systems and species less dependent on feed and other inputs can also contribute towards green growth and in this regard integrated multi-trophic systems is a promising avenue.

Aquaculture can provide beneficial jobs in rural areas with better growth potential than capture fisheries. The challenge for policy is to realise the potential of the sector while addressing the concerns that are a normal part of a new and rapidly growing industry (Box 3.12). The framework described in the OECD Green Growth Strategy can help by pointing out how to put in place innovation-friendly governance, including market-based approaches that let the private sector take the lead in addressing environmental externalities.

Key findings

Several key findings can be drawn from the discussion in this study.

- ***Green growth is feasible in aquaculture.*** There are advanced technologies, best management practices, regulations and governances available. Also there is substantial scope for green growth in aquaculture in that there are significant variations in environmental impacts between regions, production modes and species as well as in productivity.
- ***Green growth is important in aquaculture not only to increase production*** while maintaining environmental quality, ***but also to make it more acceptable to the public*** and thus increase its competitiveness in the market.
- ***Green growth in aquaculture is an on-going process***, not an end, and in this regard developed countries and developing countries can learn from each other's experience. Many developing countries have already adopted green growth policies in various forms.
- ***Sharing best practices in rearing process or policies or institutional arrangements internationally*** can help both individual farms and countries moving toward green growth. Special assistance might be appropriate in this regard for developing countries.
- ***Adopting green growth policies affects economic, social and environmental outcomes as well as the competitiveness of aquaculture secto***r. In some cases adopting a green growth agenda will lead to lower production costs and enhance competitiveness.
- ***Governments play an important role in the promotion of green growth in aquaculture*** by providing a predictable working environment, delivering innovation through R&D, introducing biosecurity measures and by setting incentives to produce within acceptable norms, etc.
- ***Environmental externalities and space competition are key issues*** to be addressed by aquaculture policy makers to ensure sustainable growth of aquaculture.
- ***Improved regulation can lead to growth.*** Effectively addressing externalities is key to unlocking future growth potential, especially in OECD countries.
- ***The co-culture of bi-valves can be effective in reducing nitrogen and phosphorus discharges.*** Market development initiatives for the produced bi-valves can help spur co-culturing, as can finding a way for producers to capture the value of their water cleaning effects.
- ***Innovation in technology*** in all aspects of aquaculture such as domestication of new species, vaccine, feed ingredients, feeding, and rearing material and system ***can help green growth at the production level while addressing the environmental challenges.***
- ***Innovations in institution and governance*** can bring needed flexibility and adaptability to management systems.
- ***Involving of all stakeholders*** can yield more effective and informed policies, improve implementation and create incentives for self- enforcement. Introducing appropriate biosecurity measures is the key safeguard to sustainable growth in aquaculture given the trend of globalisation and the increase in intensive aquaculture. The introduction of non-native species should only be done with utmost vigilance.
- ***Spatial planning*** such as integrated coastal zone management can help identify suitable farming sites, reduce uncertainty, facilitate investment, take into account environmental concerns, and help avoid conflicts among users.
- ***Introducing green growth policies*** may increase cost to producers in the short run but ***can be turned into a competitive advantage in the long run***, notably by informing consumers through

certification schemes or labelling. In this respect both public authorities and private producers have an important role to play in improving the image of the aquaculture industry.

- ***For government and international organisations*** dealing with aquaculture development ***it is imperative that the quality of statistics be improved.*** This concerns both reliability of data and coverage.
- In parallel with improved statistics, ***effective monitoring and evaluation frameworks*** needs to be developed to ensure that actions are taken and that they lead to concrete benefits. This will also help other countries still considering green growth in aquaculture (sharing experience) and can help create best practice.

Note

1. According to AquaGen, "In the last 40 years the progress in selective breeding has contributed to:
    - A reduction in production time from smolt to harvest size from 24 to 14 months
    - More efficient use of feed in that less feed is used per kilo meat produced
    - Higher survival rate, for example, resistance to the viral disease infectious pancreatic necrosis (IPN) has increased
    - Better filet quality in the areas of fat and color".

References

Allison, E.H. (2011), *Aquaculture, Fisheries, Poverty and Food Security*, (OECD internal document).

Ayer, N.W. and P.H. Tyedmers (2008), "Assessing alternative aquaculture technologies: Life cycle assessment of salmonid culture systems in Canada", *Journal of Cleaner Production*, doi:10.1016/j.jclepro.2008.08.002.

Asche, F. (2008), "Farming the Sea", *Marine Resource Economics*, Vol. 23(4): 507-527.

Asche, F., H. Hansen, and R. Tveteras (2010), "Thalassorama-The Salmon Disease Crisis in Chile", *Marine Resource Economics,* Vol. 24, pp. 405-411.

Asche, F. (2011), *Green Growth in Fisheries and Aquaculture*, OECD internal document, Paris.

Bostock, J., B. McAndrew et al. (2010), "Aquaculture: Global Status and Trends", *Philosophical transactions of the Royal Society,* Vol. 365: 2897-2912.

Burkholder, J.M. and S.E. Shumway (2011), "Bivalve shellfish aquaculture and eutrophication", in S.E. *Shumway* (ed.).

Charles, A. (2011), "Integrated Ocean Management and the Fisheries Sector: Interactions, Economic Tools and Governance Structures", OECD internal document, Paris.

Chilean Government, Undersecretariat for Fisheries and Aquaculture (2012), "The Recovery of the Chilean Salmon Industry", OECD internal document, Paris.

Coase, R. (1960), "The Problem of Social Cost", *Journal of Law and Economics*, Vol.3: 1-44.

European Commission (2013), "Strategic guidelines for the sustainable development of EU aquaculture", (COM(2013)229/final), http://ec.europa.eu/fisheries/cfp/aquaculture/official_documents/com_2013_229_en.pdf.

FAO (2014), *Fishery and Aquaculture Statistics. Global production by production source 1950-2012* (FishstatJ) in FAO Fisheries and Aquaculture Department [online or CD-ROM]. Rome. Updated 2014. http://www.fao.org/fishery/statistics/software/fishstatj/en.

FAO (2008) *Glossary of Aquaculture*, (accessed on 1 June 2011), FAO Publications, Rome. http://www.fao.org/fi/glossary/aquaculture/default.asp)

FAO (2010), *The State of World Fisheries and Aquaculture* 2010, FAO Publications, Rome. http://www.fao.org/docrep/013/i1820e/i1820e00.htm

FAO (2011), Global Statistical Collections, FAO Publications, Rome. http://www.fao.org/fishery/topic/16140/en

FAO (2011), *The role of women in agriculture*, prepared by SOFA Team 2 and Cheryl Doss, FAO Publications, Rome

Ferreira, J.G., Hawkins, A.J., Newton, T. Nickell, R. Pastres, J. Forte, A. Bodoy, and S.B. Bricker (2009), "Analysis of coastal and offshore aquaculture: application of the FARM™ model to multiple systems and shellfish species", *Aquaculture*, Vol. 289:32-41.

Fredheim, A., O. Jensen and T. Demster (2010), "OECD/ (2010), "Norway: Escapes of fish from aquaculture", in OECD, *Advancing the Aquaculture Agenda: Workshop Proceedings*, OECD Publishing, Paris.
DOI: http://dx.doi.org/10.1787/9789264088726-16-en.

Fukuyama, F., (2013), "What is Governance?", *Working Paper 314*, Center for Global Development, Washington, D.C., http://www.cgdev.org/sites/default/files/1426906_file_Fukuyama_What_Is_Governance.pdf.

Garcia, S.M. and A.A. Rosenberg (2010), "Food Security and Marine Capture Fisheries: Characteristics, Trends, Drivers and Future Perspectives", *Philosophical Transactions of the Royal Society,* Vol. 365: 2869-2880.

Gren, I.-M. et al. (2009), "Values of mussel farming for combating eutrophication: An application to the Baltic Sea", *Ecological Engineering*.

Gronroos, J., J. Seppala, F. Silvenius and T. Makinen (2006), "Life cycle assessment of Finnish cultivated rainbow trout", *Boreal Environment Research*, Vol. 11: 401-414.

G. Haraldsson (2012), "Green growth perspective on fisheries governance", Consultant Report, OECD internal document, Paris.

Hall, J.S., A. Delaporte, M.J. Phillips, M. Beveridge and M. O'Keefe (2011), *Blue Frontiers: Managing the Environmental Costs of Aquaculture.* The WorldFish Center, Penang, Malaysia

Hasan, M. and M. Halwart (2009), *Fish as feed inputs for aquaculture: Practice sustainability and implications,* FAO Fisheries and Aquaculture Technical Paper No. 518, Rome.

Jarlbæk, H. and T. Børresen. (2012), "Innovation in Fisheries and Aquaculture Technology as Contributor to a Green Growth Strategy", OECD internal document, Paris.

Jensen, O., T. Dempster, E.B. Thorstad, I. Ulem, A. Fredheim (2010), "Escapees of Fishes from Norwegian Sea-cage Aquaculture: Causes, Consequences and Prevention", *Aquaculture Environment Interactions,* Vol. 1:71-83.

Kristofersson, D. and J.L. Anderson (2006), "Is there a relationship between fisheries and farming? Inter-dependence of fisheries, animal production and aquaculture." *Marine Policy*, Vol. 30: 721–725.

Lindahl, O. and S. Kollberg (2009), "Can the EU Agri-Environmental Aid Program be Extended into the Coastal Zone to Combat Eutrophication?", *Hydrobiologia*, Vol. 629(1):59-64.

Lindahl, O., R. Hart, B. Hernroth, S. Kollerg, L. Loo, L. Olrog, A. Rehnstam-Holm, J. Svensson, S. Svensson and U. Syversen (2005), "Improving Marine Water Quality by Mussel Farming: A Profitable Solution for Swedish Society", *Ambio*, Vol. 34.

Lutz, R.A. (ed.). (1980), *Mussel Culture and Harvest: A north American Perspective*, Elservier Scientific Publishing Company, Amsterdam.

Nielsen, M., L. Ravensbeck, and R. Nielsen (2012), "*Achieving Green Growth through Improved Regulation of Fisheries and Aquaculture*", Consultant Report, OECD internal document, Paris.

Nielsen, R. (2012), *Barriers to Sustainable Growth in the Aquaculture Sector: an Economic Analysis*, Ph.D. Thesis, University of Copenhagen, Denmark.

Norwegian Ministry of Fisheries and Coastal Affairs (2009), *Strategy for an Environmentally Sustainable Norwegian Aquaculture Industry*, Oslo.

Norwegian Ministry of Fisheries and Coastal Affairs (2012a), *Moving towards a Zoning Structure in the Norwegian Aquaculture?* Consultant Report, OECD internal document, Paris.

Norwegian Ministry of Fisheries and Coastal Affairs (2012b), *Innovations Strategies and Green Growth-Norwegian Salmon Farming Case*, OECD internal document, Paris.

OECD/FAO (2013), *OECD-FAO Agricultural Outlook 2013*, OECD Publishing, Paris. DOI: http://dx.doi.org/10.1787/agr_outlook-2013-en

OECD/FAO (2012a), *OECD-FAO Agricultural Outlook 2012*, OECD Publishing, Paris. DOI: http://dx.doi.org/10.1787/agr_outlook-2012-en.

OECD (2012b), "Green growth and aquaculture Workshop held on 12-13 December 2012, Yeosu, Korea Summary of Discussions/Presentations and Case Studies", OECD internal document, Paris.

OECD (2011a), *Towards Green Growth: Green Growth Strategy Synthesis Report*, OECD internal document, Paris.

OECD (2011b), *Green Growth Strategy Synthesis Report*, Consultant Report, OECD internal document.

OECD (2011c), "Summary Record of the 108th Session of the Committee for Fisheries", OECD internal document, Paris.

OECD (2011d), *Towards Green Growth: Monitoring Progress: OECD Indicators*, OECD Green Growth Studies, OECD Publishing, Paris.
DOI: http://dx.doi.org/10.1787/9789264111356-en.

OECD (2010a), *Advancing the Aquaculture Agenda: Workshop Proceedings*, OECD Publishing, Paris.
DOI: http://dx.doi.org/10.1787/9789264088726-en.

OECD (2010b), *Conditions for Establishing Aquaculture Production Sites in OECD Countries*, (OECD internal document).

OECD/R. Díaz (2010c), "Agriculture's impact on aquaculture: Hypoxia and eutrophication in marine waters", in OECD, *Advancing the Aquaculture Agenda: Workshop Proceedings*, OECD Publishing, Paris.
DOI: http://dx.doi.org/10.1787/9789264088726-20-en

Ostrom, E. (1990). *Governing the Commons: The Evolution of Institutions for Collective Action.* Cambridge University Press, UK.

Pelletier, N., P. Tyedmers, U. Sonesson, A. Scholz, F. Ziegler, A. Flysjo, S. Kruse, B. Cancino, and H. Silverman (2009), "Not all salmon are created equal: life cycle assessment (LCA) of global salmon farming systems", *Environmental Science & Technology*, Vol. 43, No. 23.

Peterson, J.K., and L-O. Loo (2004). Miljokonsekvenser of dirking av blamuslinger. Interreg project III-A nr. GS 3041-45-02 "Blaskjellanlegg og nitrogenkvoter," Final Report 31.8.2004 (In Danish) In S.E. *Shumway* (ed.).

Selman, M., S. Greenhalgh, R. Díaz and Z. Sugg (2008), "Eutrophication and Hypoxia in Coastal Areas: A Global Assessment of the State of Knowledge", *World Resources Institute Policy Note*, No. 1, Washington D.C., pp. 1-6.

Sen, S. (2001), "Involving Stakeholders in Aquaculture Policy-making, Planning and Management", *Technical Proceedings of the Conference on Aquaculture in the Third Millennium*, Bangkok, Thailand, 20-25 February 2000, NACA, Bangkok and FAO, Rome.

de Serres, A., F. Murtin and G. Nicoletti (2010), "A Framework for Assessing Green Growth Policies", *OECD Economics Department Working Papers*, No. 774, OECD Publishing, Paris.
DOI: http://dx.doi.org/10.1787/5kmfj2xvcmkf-en.

Shumway, S.E. (ed.) (2011), *Shellfish Aquaculture and the Environment,* Wiley-Blackwell, Singapore.

Tacon, A.G.J., and M. Metian (2008), "Global Overview on the Use of Fish meal and Fish oil in Industrially Compounded Aquafeeds: Trends and Future Prospects", *Aquaculture*, Vol. 285: 146-158.

Tacon, A.G.J., and M. Metian (2009), "Fishing for Aquaculture: Non-food use of Small Pelagic Forage Fish-a Global Perspective", *Reviews of Fisheries Science*, Vol. 17: 305-317.

Welch, A. et al.(2010), "From Fishing to the Sustainable Farming of Carnivorous Marine Finfish", *Reviews in Fisheries Science*, Vol. 18, No. 3.

World Fish Center (2009), "Gleaner, fisher, trader, processor: Understanding gendered employment in the fisheries and aquaculture sector", paper by N. Weeratunge and C. Snyder at the FAO-IFAD-ILO Workshop on Gaps, trends and current research in gender dimensions of agricultural and rural employment: Differentiated pathways out of poverty, Rome, 31 March – 2 April.

ORGANISATION FOR ECONOMIC CO-OPERATION AND DEVELOPMENT

The OECD is a unique forum where governments work together to address the economic, social and environmental challenges of globalisation. The OECD is also at the forefront of efforts to understand and to help governments respond to new developments and concerns, such as corporate governance, the information economy and the challenges of an ageing population. The Organisation provides a setting where governments can compare policy experiences, seek answers to common problems, identify good practice and work to co-ordinate domestic and international policies.

The OECD member countries are: Australia, Austria, Belgium, Canada, Chile, the Czech Republic, Denmark, Estonia, Finland, France, Germany, Greece, Hungary, Iceland, Ireland, Israel, Italy, Japan, Korea, Luxembourg, Mexico, the Netherlands, New Zealand, Norway, Poland, Portugal, the Slovak Republic, Slovenia, Spain, Sweden, Switzerland, Turkey, the United Kingdom and the United States. The European Commission takes part in the work of the OECD.

OECD Publishing disseminates widely the results of the Organisation's statistics gathering and research on economic, social and environmental issues, as well as the conventions, guidelines and standards agreed by its members.

OECD PUBLISHING, 2, rue André-Pascal, 75775 PARIS CEDEX 16
(51 2015 03 1 P) ISBN 978-92-64-23213-6 – 2015

www.ingramcontent.com/pod-product-compliance
Ingram Content Group UK Ltd.
Pitfield, Milton Keynes, MK11 3LW, UK
UKHW051507290926
15471UKWH00008B/1465

9 789264 232136